Solution Manual for:

100 Genesys Design Examples

Second Edition

Based on the Textbook:

Microwave and RF Engineering

Solution Manual for:

100 GENESYS Design Examples

Second Edition

Based on the Textbook:
Microwave and RF Engineering

Ali Behagi

Techno Search

Ladera Ranch, CA 92694

Solution Manual for:

100 Genesys Design Examples

Second Edition

Based on the Textbook:

Microwave and RF Engineering

Copyright © 2019 by Techno Search

ISBN: 978-0-9835460-4-7

First Published in USA

Techno Search

Ladera Ranch, CA 92694

Table of Contents

Chapter 3 Network Parameters and the Smith Chart   57

Chapter 4 Resonant Circuits and Filter Design   73

Chapter 5 Power Transfer and Impedance Matching

Chapter 8 Multi-Stage Amplifier Design   225

Table of Examples

Chapter 3: Network Parameters and the Smith Chart   57

Chapter 8: Multi-Stage Amplifier Design   225

Foreword

The Second Edition of the 100 Genesys Design Examples book consolidates relevant knowledge and practical skills that are highly sought-after in the RF and microwave industry. This book provides practical hands-on experience for the practicing engineers or university students to quickly acquire the practical understanding of RF and microwave circuit design. This is made possible by the well-chosen design examples and using the Keysight Genesys software for their solution. The powerful synthesis and simulation tools in Genesys software are used by more than 5,000 RF and microwave engineers worldwide.

Prof. Behagi has thoughtfully incorporated theoretical RF and microwave design concepts through familiar MATLAB scripting and equations in the Genesys simulation examples so that readers can use them for learning or as starting points in their designs. Users of Genesys can interactively change design parameters and see results instantly through convenient simulations that have already been set up in these examples. Prof. Behagi's innovative teaching approach breaks the barrier between theory and practice and allows the knowledge learned in the engineering schools to be applied effectively on the job.

By judiciously selecting an industrial strength RF and microwave simulation tool that is also easy to learn and use for the teaching of RF and microwave engineering topics, Prof. Behagi has transformed the learning of this difficult subject from being intimidating to being enjoyable. It is also personally fulfilling because the reader will acquire skills highly valued in the job market by companies designing RF and microwave products, which range from consumer wireless devices to advanced aerospace and defense systems.

I believe the reader will enjoy learning and using the knowledge contained in this book in conjunction with Keysight Genesys sofware for productive RF and microwave circuit designs.

How-Siang Yap
Keysight Genesys Planner and Product Manager
Keysight EEsof EDA
Keysight Technologies Inc.

Preface

The Second Edition of the 100 Genesys Design Examples book is mainly written for university students and practicing engineers who know the basic theory of analog RF and microwave engineering and want to apply the theory to the analysis and design of practical RF and microwave circuit design using the Keysight Genesys software. The 100 examples are taken from the Microwave and RF Engineering textbook written be same author.

The 100 Genesys Design Examples book is divided into eight chapters.

1. RF and Microwave Components

2. Transmission Line Components

3. Network Parameters and the Smith Chart

4. Resonant Circuits and Filter Design

5. Power Transfer and Impedance Matching

6. Distributed Impedance Matching

7. Single Stage Amplifier Design

8. Multi-Stage Amplifier Design

The 100 Genesys Design Examples book has an associated 100 Genesys Workspaces that comes with the book. University students and practicing engineers will find the book both as a potent learning tool and as a reference guide to quickly setup designs using the Genesys software. The author also uses CAD techniques that may not be familiar to some engineers. This includes subjects such as the frequent use of the MATLAB scripting capability.

The 100 Genesys Workspaces can be found at the following URL.

http://www.keysight.com/find/eesof-genesys-rfmw-workspaces

Chapter 1

RF and Microwave Components

1.1 Straight Wire Inductance

A conducting wire carrying an AC current produces a changing magnetic field around the wire. According to Faraday's law the changing magnetic field induces a voltage in the wire that opposes any change in the current flow. This opposition to change is called self inductance. At high frequencies even a short piece of straight wire possesses frequency dependent resistance and inductance behaving as a circuit element.

1.2 Analysis of Straight Wire in Genesys

Example 1-1: Calculate the reactance and inductance of a three inch length of AWG #28 copper wire in free space at 60 Hz, 500 Hz, and 1 GHz.

Solution: To analyze the Example in Genesys, create a schematic and add the straight wire model from the Parts Library, as shown in Figure 1-1.

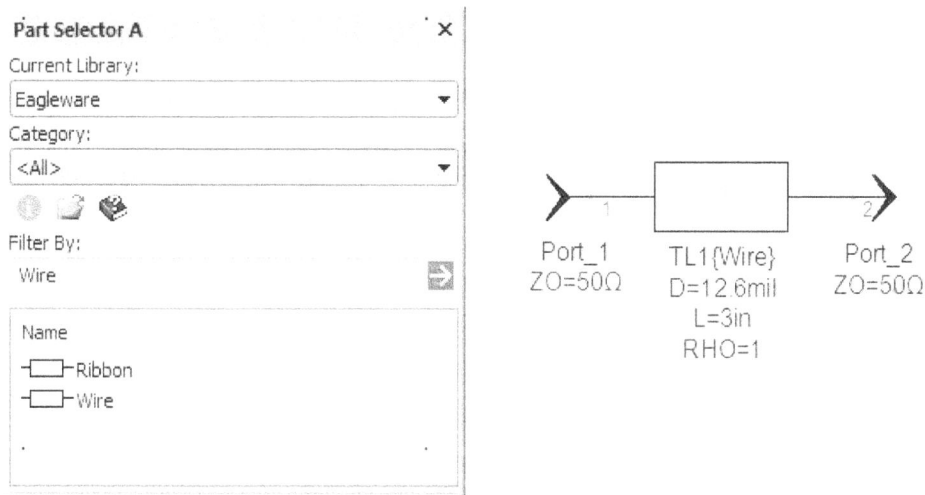

Figure 1-1 Part Selector and schematic of the straight wire

Attach 50 Ohm input and output ports, set the wire diameter to 12.6 mils, wire length to 3 inches and Rho=1, as shown in Figure 1-1. Rho is not the actual resistivity of the wire but rather the resistivity of the wire relative to copper. Because we are modeling a copper wire the value should be set to one. Table 1-1 provides a reference of common materials in terms of their actual resistivity and the relative resistivity to copper.

Material	Resistivity Relative to Copper	Actual Resistivity Ω-meters	Actual Resistivity Ω-inches
Copper, annealed	1.00	$1.68 \cdot 10^{-8}$	$6.61 \cdot 10^{-7}$
Silver	0.95	$1.59 \cdot 10^{-8}$	$6.26 \cdot 10^{-7}$
Gold	1.42	$2.35 \cdot 10^{-8}$	$9.25 \cdot 10^{-7}$
Aluminum	1.64	$2.65 \cdot 10^{-8}$	$1.04 \cdot 10^{-6}$
Tungsten	3.25	$5.60 \cdot 10^{-8}$	$2.20 \cdot 10^{-6}$
Zinc	3.40	$5.90 \cdot 10^{-8}$	$2.32 \cdot 10^{-6}$
Nickel	5.05	$6.84 \cdot 10^{-8}$	$2.69 \cdot 10^{-6}$
Iron	5.45	$1.00 \cdot 10^{-7}$	$3.94 \cdot 10^{-6}$
Platinum	6.16	$1.06 \cdot 10^{-7}$	$4.17 \cdot 10^{-6}$
Tin	52.8	$1.09 \cdot 10^{-7}$	$4.29 \cdot 10^{-6}$
Nichrome	65.5	$1.10 \cdot 10^{-6}$	$4.33 \cdot 10^{-5}$

Table 1-1 Resistivity of common materials relative to copper

Create a Linear analysis to analyze the circuit's impedance versus frequency. The Linear Analysis Properties window is shown in Figure 1-2.

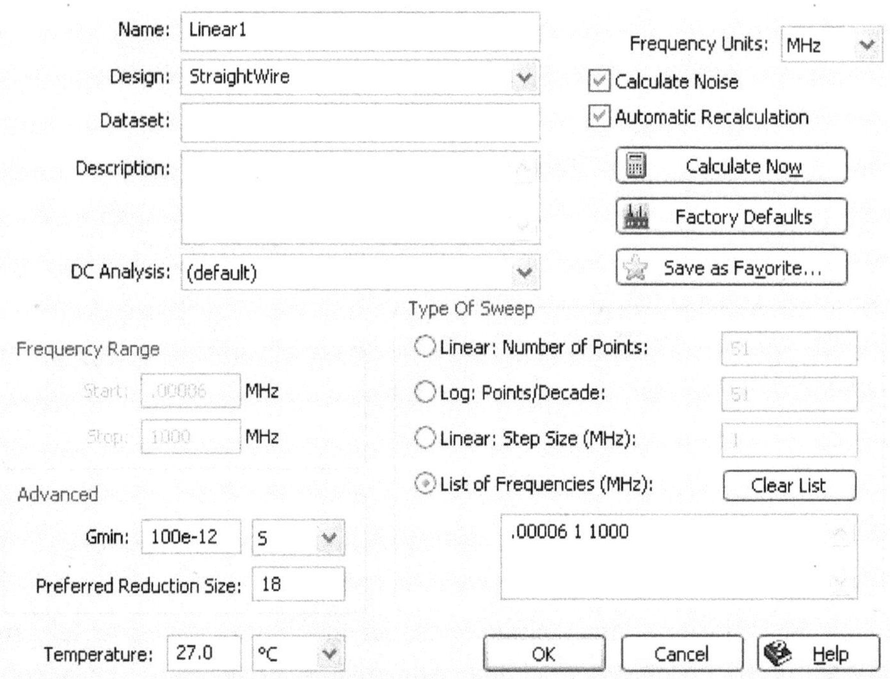

Figure 1-2 Circuit analysis properties window

Create a list of frequencies under the *Type of Sweep* setting. Enter the frequencies of 60 Hz, 1 MHz, 1 GHz. When an analysis is run the results are written to a Dataset. The results of a Dataset may then be sent to a graph or tabular output for visualization. In this example an Equation Editor is used to post process the solutions in the Dataset. Create an Equation Editor and type equations for the calculation of the wire's reactance and inductance, as shown in Figure 1-3.

```
1    using("Linear1_Data")
2    reactance=im(Linear1_Data.ZIN1)
3  . inductance=reactance./(F*2*PI)
```

Figure 1-3 MATLAB equations for calculation of Inductance

More complex workspaces may contain multiple Datasets. It is a good practice to specify which Dataset is used to collect data for post processing. This is accomplished with the {using ("Linear1_Data")} statement of line 1 in the Equation Editor. Genesys has a built-in function ZIN1 to calculate the impedance of the circuit at each analysis frequency. Line 2 defines the reactance as the imaginary part of the impedance. Line 3 calculates the inductance from the reactance using the equation $L = X/2*PI*F$. The frequency, F, is the independent variable created by the Linear Analysis. Note the use of the dot (.) notation in the Equation Editor. There is a dot after Linear1_Data and reactance of lines 2 and 3. The dot means that these variables are not singular quantities but are arrays of values. There is a calculated array value for each independent variable, F. The Equation Editor is an extremely powerful feature of the Genesys software and is used frequently throughout this text. It is an interactive mathematical processor similar to MATLAB by MathWorks. There are two different syntaxes that may be used to define equations and perform post processing operations using the Equation Editor: The Engineering Language and the MATLAB Script. The Engineering Language is a simple structured format as shown in Figure 1-3. The MATLAB Script is compatible with the m-file syntax that is used in MATLAB. It is very convenient for students and engineers that are proficient in MATLAB. Both types will be used throughout this book to demonstrate the use of both languages. Under Equation command either Engineering Language or MATLAB Script can be selected. The simulated value of variables can be displayed in a Workspace Variables window or sent to a tabular output. Figure 1-4 shows the Workspace Variables at 3 different frequencies and the same variables in tabular output.

Name	Value
inductance (3)	[92.89e-9; 92.19e-9; 89.42e-9]
reactance (3)	[35.02e-6; 0.579; 561.863]

Table2			
Index	F (MHz)	reactance	inductance
1	60e-6	35.02e-6	92.89e-9
2	1	0.579	92.19e-9
3	1000	561.863	89.42e-9

Figure 1-4 Workspace Variables window and output Table

We can see that at 60 Hz and 1 MHz the reactance and resulting inductance are less than 1 Ohm . At 1 GHz however, the values of reactance is greater than 560 Ohm. In general as the frequency enters the RF and microwave region, the value of reactance begins to increase. This is normal and due to the skin effect of the conductor. The skin effect is a property of conductors where, as the frequency increases, the current density concentrates on the outer surface of the conductor due to a decrease in the surface area.

1.3 Calculation of Flat Ribbon Inductance

Flat ribbon style conductors are very common in RF and microwave engineering. Flat ribbon conductors are encountered in RF systems in the form of low inductance ground straps. Flat ribbon conductors are encountered in microwave integrated circuits (MIC) as gold bonding straps. When a very low inductance is required the flat ribbon or copper strap is a good choice.

The flat ribbon inductance can be calculated from the following empirical equation.

$$L = K \ell \left[\ln\left(\frac{2\ell}{W+T}\right) + 0.223\left(\frac{W+T}{\ell}\right) + 0.5 \right] \quad nH$$

where:
    ℓ = The length of the wire
    K= 2 for dimensions in cm and K=5.08 for dimensions in inches
    W = the width of the conductor
    T = the thickness of the conductor

Example 1-2: Calculate the inductance of the 3 inch Ribbon at 60 Hz, 1 MHz, and 1 GHz. Make the ribbon 100 mils wide and 2 mils thick.

Solution: The schematic of the Ribbon in Genesys is shown in Figure 1-5 along with the table with the reactance and inductance. Note that the same length of ribbon as the AWG#28 wire has almost half of the inductance value. Also note that the inductance value, 49.69 nH, does not change over the 60 Hz to 1 GHz frequency range. It is why the ribbon is desirable as a low inductance conductor for use in RF and microwave applications.

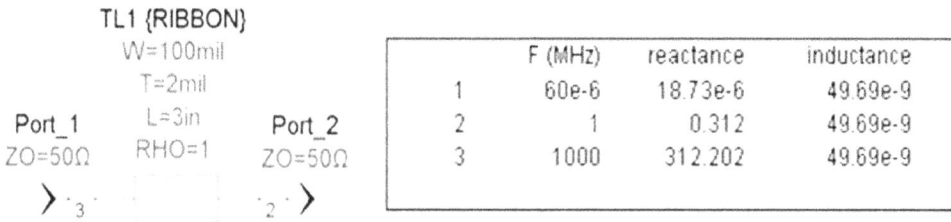

TL1 {RIBBON}
W=100mil
T=2mil
L=3in
RHO=1

	F (MHz)	reactance	inductance
1	60e-6	18.73e-6	49.69e-9
2	1	0.312	49.69e-9
3	1000	312.202	49.69e-9

Port_1 ZO=50Ω   Port_2 ZO=50Ω

Figure 1-5 Flat ribbon schematic and tabular output

1.4 Ideal and Physical Resistors

The resistance of a material determines the rate at which electrical energy is converted to heat. In Table 1-1 we have seen that the resistivity of material is specified in Ω-meters rather than Ω/meter. This facilitates the calculation of resistance. At low frequency or logic circuits we often treat resistors as ideal components.

Example 1-3: Plot the impedance of a 50 Ω ideal resistor in Genesys over a frequency range of 0 to 2 GHz.

 Solution: For the 50 Ω ideal resistor the plot is shown in Figure 1-6. This plot shows a constant resistance at all frequencies.

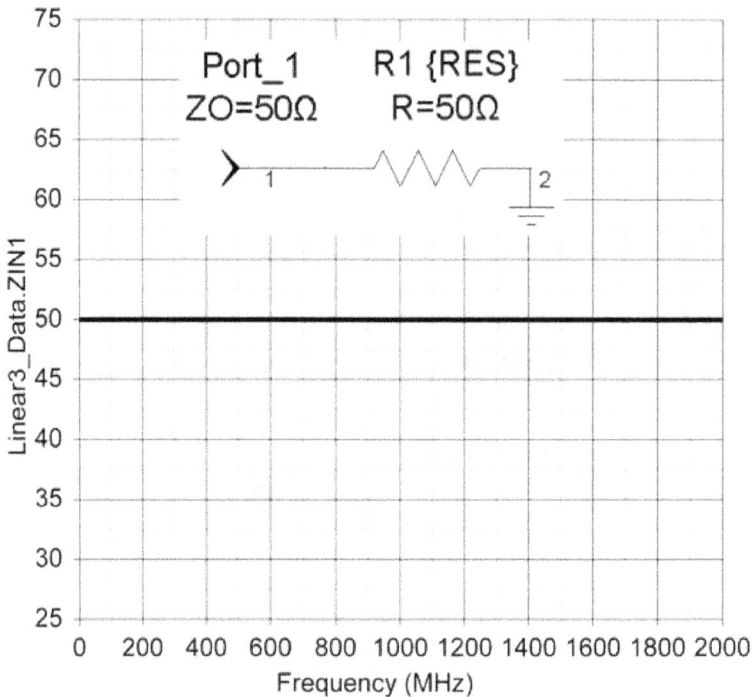

Figure 1-6 Ideal 50Ω resistor impedance versus frequency

Example 1-4: Plot the impedance of a 50 Ohm leaded resistor in Genesys over a frequency range of 0 to 2 GHz.

Solution: At RF and microwave frequencies however, leaded resistors also possess inductive and capacitive elements. The stray inductance and capacitance associated with leaded resistors are often called parasitic elements. Consider the leaded resistor as shown in Figure 1-7. For such a 1/8 watt leaded resistor it is not uncommon for each lead to have about 10 nH of inductance. The body of the resistor may exhibit 0.5 pF capacitance between the leads. Designing the network in Genesys reveals an interesting result of the impedance versus frequency response. The impedance plotted in Figure 1-7 shows the impedance for a 10 Ω, 50 Ω, 500 Ω, and 1 kΩ leaded resistors swept from 0 to 2.0 GHz.

By tuning the parasitic elements in Genesys we find that the low value resistors are influenced more by the lead inductance. The high value resistors are influenced more by the parasitic capacitance.

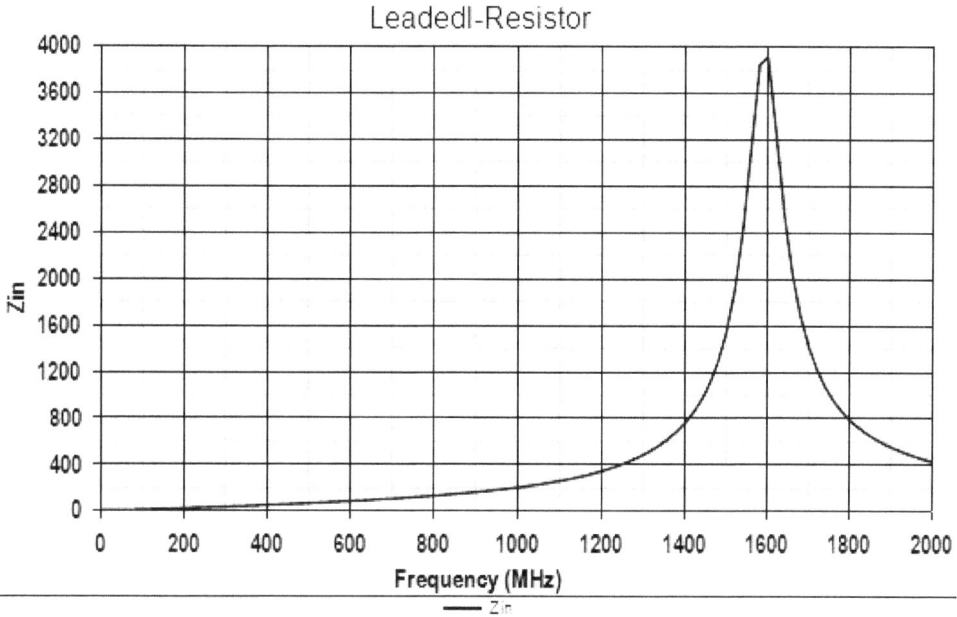

Figure 1-7 Leaded resistor impedance versus frequency

1.5 Chip Resistors

Thick film resistors are used in most contemporary electronic equipment. The thick film resistor, often called chip resistor, comes close to eliminating much of the inductance that plagues the leaded resistor. The chip resistor works well with popular surface mount assembly techniques preferred in modern electronic manufacturing.

Example 1-5: Plot the impedance of 1 kΩ, 0603 size chip resistor, manufactured by KOA, from 0 to 3 GHz.

Solution: The KOA resistor model is shown in the Modelithics Library.

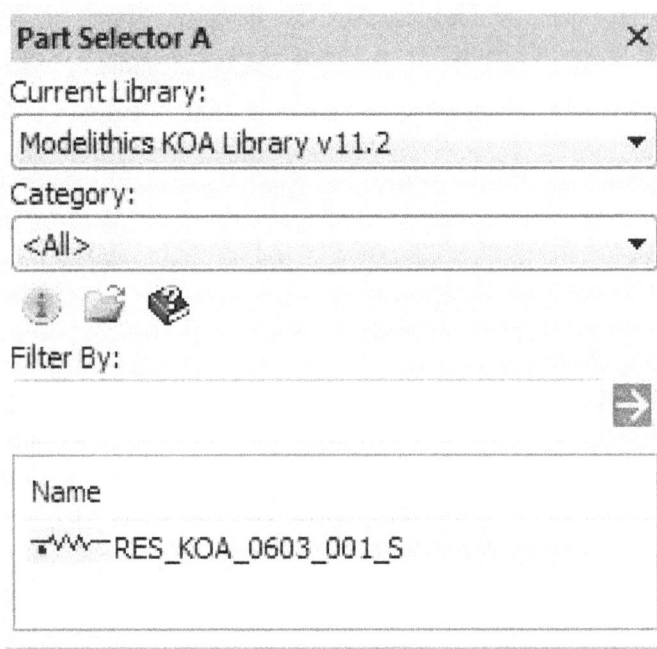

Figure 1-8 Modelithics SELECT library in Genesys

Solution: Create a schematic with the resistor and sweep the impedance from 0 to 3 GHz. The resulting schematic and response are shown in Figure 1-9.

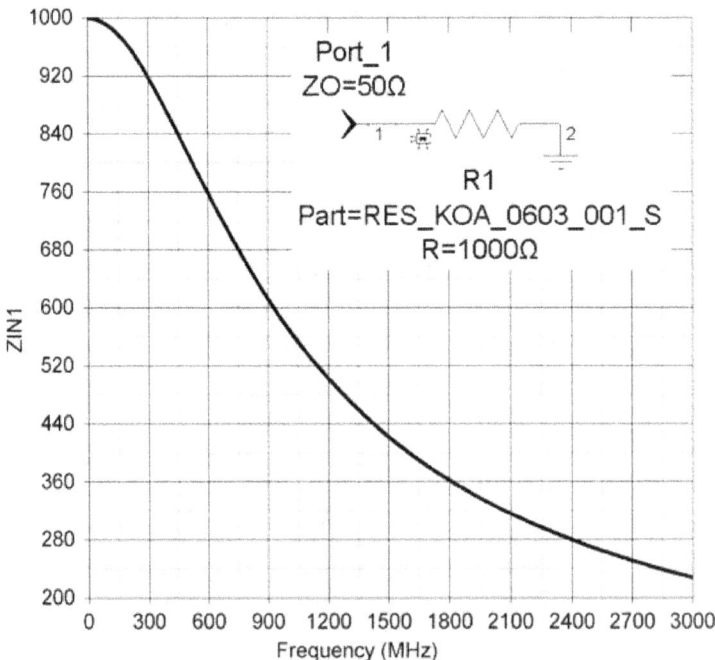

Figure 1-9 Modelithics chip resistor model schematic and impedance versus frequency

Note the roll off of the impedance with increasing frequency. This suggests that the chip resistor does have a parasitic capacitance that is in parallel with the resistor.

1.6 Air Core Inductors

Forming a wire on a removable cylinder is the basic realization of the air core inductor. When designing an air-core inductor, use the largest wire size and close spaced windings to result in the lowest series resistance and high Q. The basic empirical equation to calculate the inductance of an air core inductor is given by the following equation.

$$L = \frac{(17)N^{1.3}(D+D1)^{1.7}}{(D1+S)^{0.7}}$$

Where:

N = Number of turns of wire

D = Core form diameter in inches

$D1$ = Wire diameter in inches

L = Coil inductance in nH

S = Spacing between turns in inches

Example 1-6: Calculate the amount of inductance that we can realize in that same three inches of wire if we wind it around a core to form an inductor. Choose a core form of 0.095 inches diameter as a convenient form to wrap the wire around.

Solution: First we need to calculate the approximate number of turns that we can expect to have with the three inch length of wire. We know that the circumference of a circle is related to the diameter by the following equation.

$$Circumference = \pi\,(0.095) = 0.2985 \quad inches$$

With a circumference of 0.2985 inches we can calculate the approximate number of turns that we can wrap around the 0.095 inch core with three inches of wire.

$$N = \frac{3}{0.2985} = approximately \; 10 \; turns$$

From the equation for inductance we can see that the spacing between the turns has a strong effect on the value of the inductance that we can expect from the coil. When hand winding the coil, it may be difficult to maintain an exact spacing of zero inches between the turns. Therefore it is useful to solve the inductance equation in terms of a variety of coil spacing so that we can see the effect on the inductance. The MATLAB equations in Figure 1-11 is used to solve the inductance equation for a variety of coil spacing.

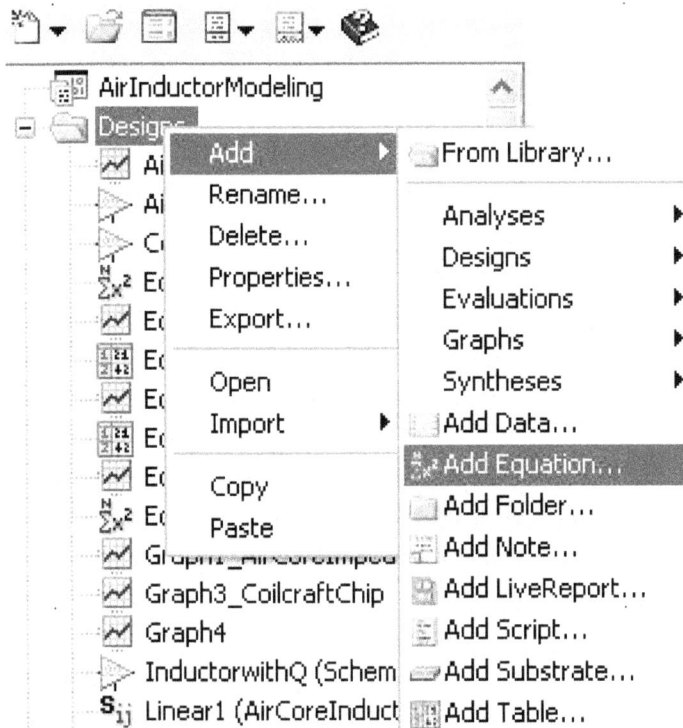

Figure 1-10 Adding Equation Editor to the Genesys workspace

Write the equation set as shown in Figure 1-11.

```
D=.095
D1=.0126
N=10
Spacing=[0;.002;.004;.006;.008;.010]
Inductance_nH=(17*(N^1.3)*((D+D1)^1.7))/((D1+Spacing)^.7)
Coil_Length=(D1*N)+(Spacing.*(N-1))
```

Figure 1-11 MATLAB equations to calculate the air core coil inductance

Note that the coil spacing variable, Spacing, has been defined as an array variable. Placing a semicolon between the values makes the array organized in a column format. This is handy for viewing the results in tabular format. Placing a comma between the values would organize the array in a row format. The array variables are displayed under the variable column as Real, indicating that these values are a six element array containing scalar values. As an aid in forming the coil, the overall coil length is also calculated within the Equation Editor. The coil length is simply the summation of the overall wire thickness times the number of turns and the spacing between the turns. A quick way to add the results to a table is shown in Figure 1-12.

Name		Value
Coil_Length	(6)	[0.126; 0.144; 0.162; 0.18; 0.198; 0.216]
D		0.095
D1		0.013
Inductance_nH	(6)	[163.784; 147.735; 135.037; 124.701; 116.097; 108.806]
N		10
Spacing	(6)	[0; 2e-3; 4e-3; 6e-3; 8e-3; 0.01]

Figure 1-12 Sending variables to a Table directly from the Equation Editor

Right click on any variable and add it to an existing or new table. Add the inductance, spacing, and coil length to the table as shown in Figure 1-13.

Index	Inductance_nH	InductanceCalculator.Spacing	Coil_Length
1	163.784	0	0.126
2	147.735	2e-3	0.144
3	135.037	4e-3	0.162
4	124.701	6e-3	0.18
5	116.097	8e-3	0.198
6	108.806	0.01	0.216

Figure 1-13 Coil Inductance versus coil spacing and length

1.7 Modeling the Air Core Inductor

Example 1-7: Calculate and plot the input impedance of an air core inductor.

Solution: The air core inductor is modeled in Genesys using the AIRIND1 model as shown in Figure 1-14.

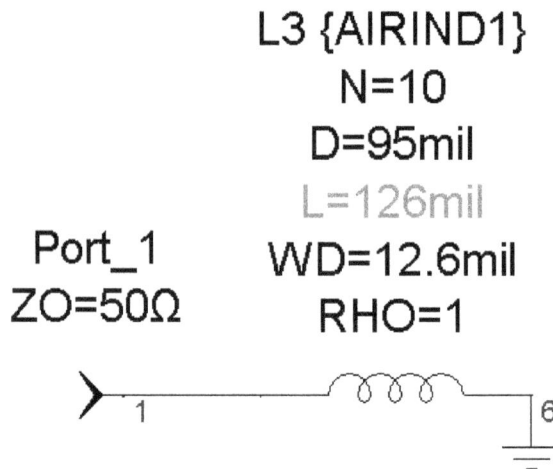

L3 {AIRIND1}
N=10
D=95mil
L=126mil

Port_1   **WD=12.6mil**
ZO=50Ω   **RHO=1**

Figure 1-14 Genesys model of the air core inductor

Note the required parameters: number of turns, wire diameter, core diameter, and coil length. The coil spacing cannot be entered directly but is accounted for by the overall coil length for a given number of turns. Make the coil length a variable so that we can analyze the inductance as a function of the coil length. Simulate the value of inductance at a fixed frequency of 1 MHz. Set up a fixed frequency Linear Analysis as shown in Figure 1-15. Simulate the value of inductance at a fixed frequency of 1 MHz.

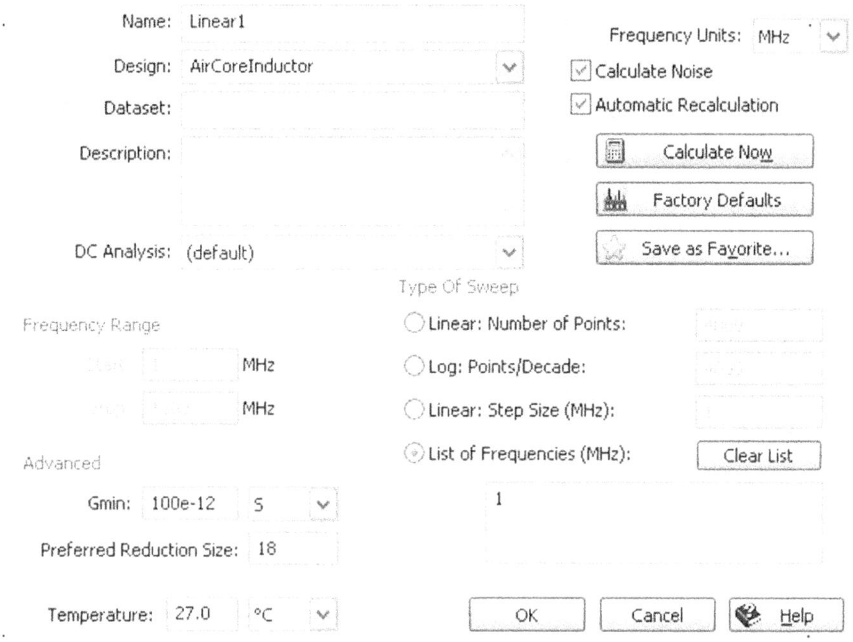

Name: Linear1

Frequency Units: MHz

Design: AirCoreInductor

☑ Calculate Noise

Dataset:

☑ Automatic Recalculation

Description:

Calculate Now

Factory Defaults

DC Analysis: (default)

Save as Favorite...

Type Of Sweep

Frequency Range

◯ Linear: Number of Points:

Start   MHz

◯ Log: Points/Decade:

Stop   MHz

◯ Linear: Step Size (MHz):

Advanced

◉ List of Frequencies (MHz):   Clear List

Gmin: 100e-12   S

1

Preferred Reduction Size: 18

Temperature: 27.0   °C

OK   Cancel   Help

Figure 1-15 Linear analysis at 1MHz

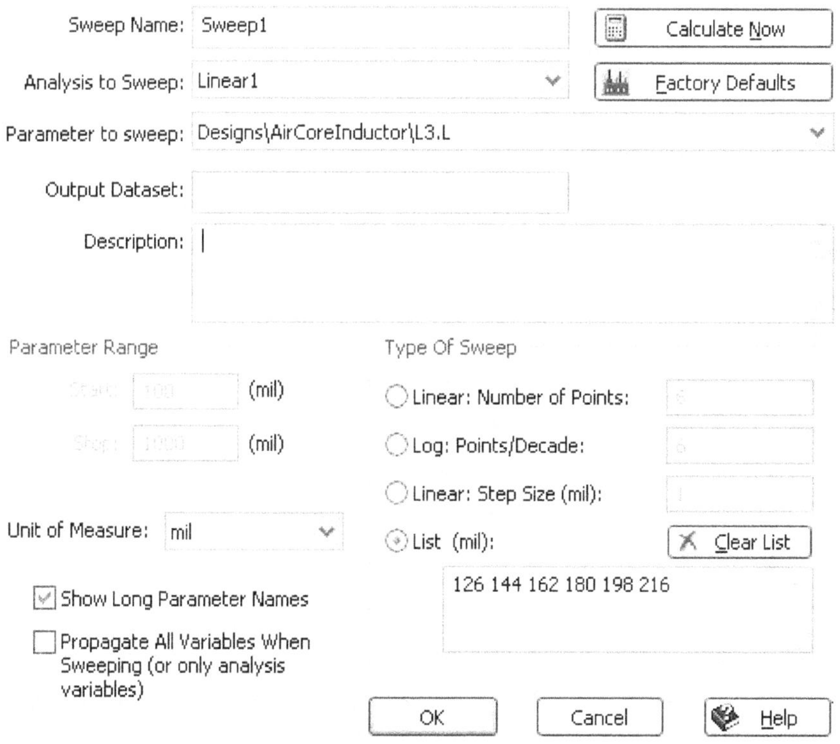

Sweep Name: Sweep1

Calculate Now

Analysis to Sweep: Linear1

Factory Defaults

Parameter to sweep: Designs\AirCoreInductor\L3.L

Output Dataset:

Description: |

Parameter Range

Type Of Sweep

Start   100   (mil)

◯ Linear: Number of Points:

Step   1000   (mil)

◯ Log: Points/Decade:

◯ Linear: Step Size (mil):

Unit of Measure: mil

◉ List (mil):   Clear List

☑ Show Long Parameter Names

126 144 162 180 198 216

☐ Propagate All Variables When
   Sweeping (or only analysis
   variables)

OK   Cancel   Help

Figure 1-16 Adding Parameter Sweep to the workspace

To vary the length of the inductor we will use the Parameter Sweep capability in Genesys. Add a Parameter Sweep to the Workspace as shown by MATLAB equations in Figure 1-17.

```
1    using("Sweep1_Data")
2    reactance=im(Sweep1_Data.ZIN1)
3    inductance=reactance./((Sweep1_Data.F)*2*PI)
4    setindep(*inductance","Sweep1_Data.L3_L_Swp_F")
```

Figure 1-17 MATLAB equations with parameter sweep dataset

On the Parameter-to-Sweep entry make sure that the coil length is selected from the drop down box. Under the Type-of-Sweep, select list and enter the six coil lengths that were calculated in the table of Figure 1-12. Finally use MATLAB equations to calculate the coil's inductance from the simulated reactance and plot as a function of the coil length as shown in Figure 1-18.

Note the use of the Setindep statement in line 4. This keyword is used to set the dependent and independent variables. The first variable in the statement is the dependent variable while the second is the independent variable.

In this example the inductance is the dependent variable while the coil length is the independent variable. As the plot of Figure 1-18 shows the model has very close correlation with the inductance calculated in the previous section.

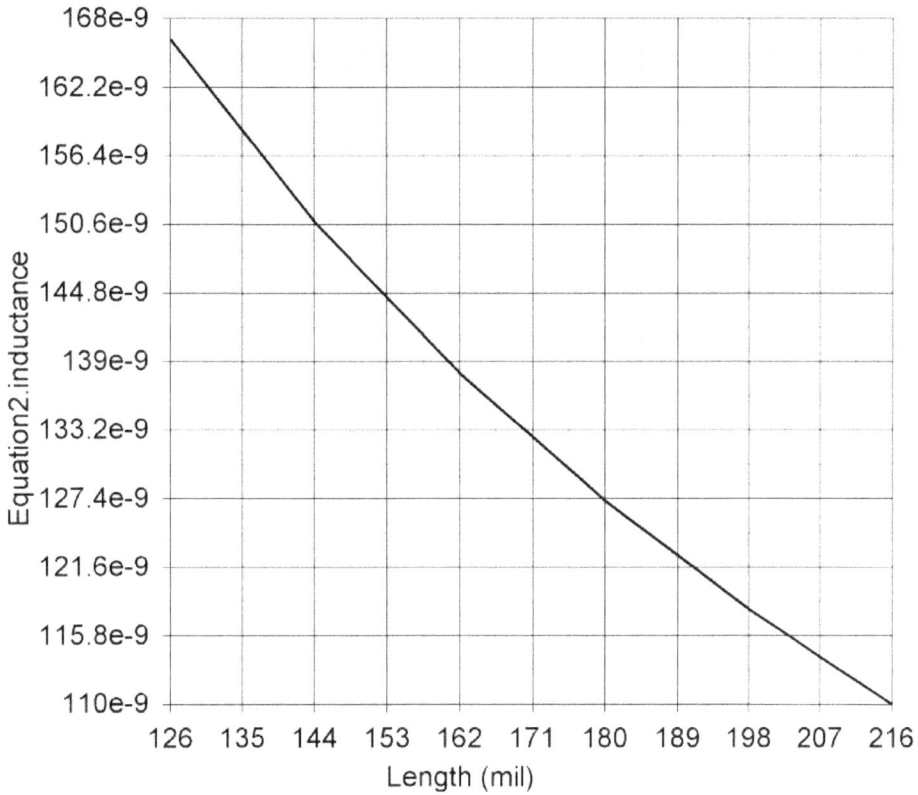

Figure 1-18 Plot of inductance versus coil length for the air inductor model

Change the Linear Analysis to a linear frequency sweep with 401 points over a range of 1 MHz to 1300 MHz. Plot the impedance of the inductor across the frequency range as shown in Figure 1-19. Note the interesting spike, or increase in impedance that occurs around 1098.6 MHz. This is the parallel, self resonant, frequency of the inductor.

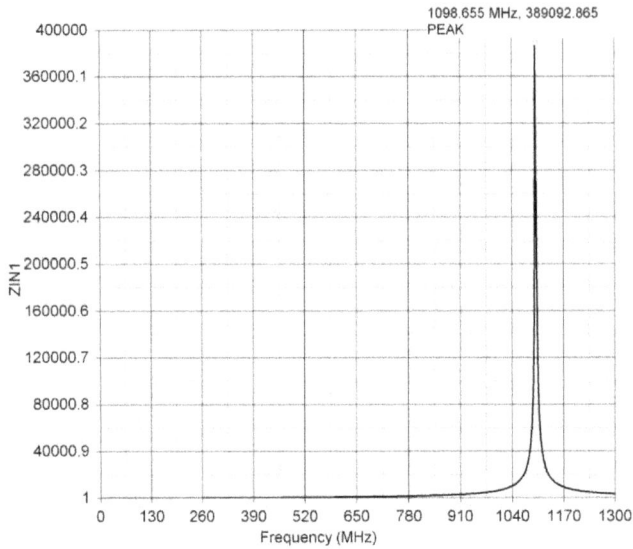

Figure 1-19 Impedance of the air core inductor as a function of frequency

Note that the inductor is not an ideal component or a pure inductance but rather a network that includes parasitic capacitance and resistance.

Example 1-8: Create a simple RLC network that gives an equivalent impedance response similar to Figure 1-19.

Solution: The schematic and display are shown in Figure 1-20.

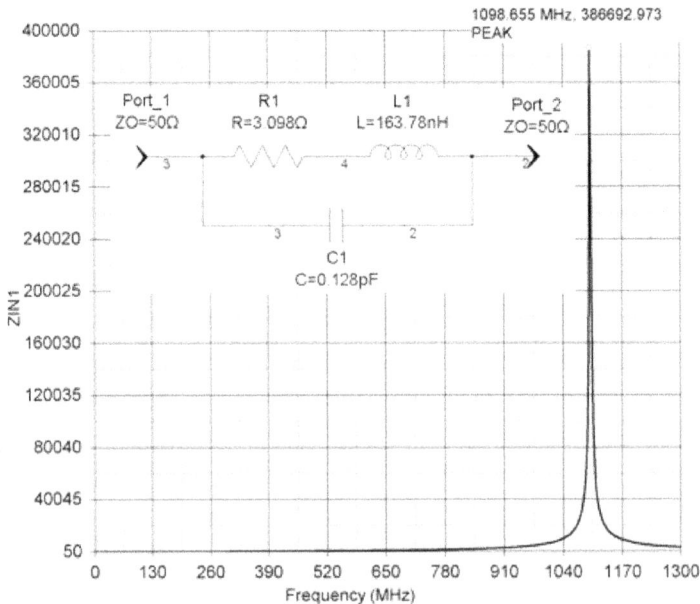

Figure 1-20 Equivalent ideal element network of the air core inductor

1.8 Inductor Q Factor

Example 1-9: Calculate the Q factor of the air core inductor given in Example 1-6. Plot the Q factor versus frequency.

Solution: The Equation Editor is shown in Figure 1-21.

```
1      using("Linear3_Data")
2      reactance=im(Linear3_Data.ZIN1)
3      inductance=abs(reactance./(F*2*PI))
4      resistance=re(Linear3_Data.ZIN1)
5      Qfactor=abs(reactance)/resistance
```

Figure 1-21 MATLAB equations to calculate the Q factor of the inductor

The plot of inductor Q factor versus frequency is shown in Figure 1-22.

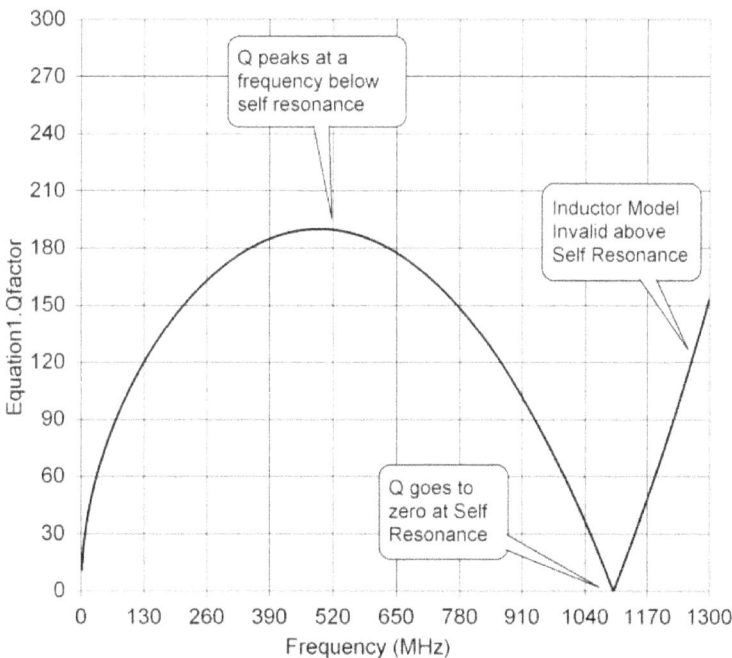

Figure 1-22 Air core inductor model Q factor versus frequency

It is interesting to note that the Q factor peaks at a frequency well below the self resonant frequency of the inductor. The actual frequency at which the Q factor peaks will vary among inductor designs but is usually ranges from 2 to 5 times less than the SRF. Close winding spacing results in inter-winding capacitance, which lowers the self-resonant frequency of the inductor. Thus there is a tradeoff between maximum Q factor and high self resonant frequency. It is also noteworthy that the Q factor goes to zero at the self resonant frequency. Figure 1-22 shows the Q factor monotonously increasing beyond the self resonant frequency. This is erroneous and is due to the fact that the model used to simulate the inductor's performance is invalid beyond the self resonance. The Air Core inductor model uses a simplified network similar to the one shown in Figure 1-20. Beyond self resonance the complexity and number of ideal elements need to increase in order to accurately model the inductor. A resistor needs to be added in series with the capacitor to begin to model the Q factor because the inductor is becoming a capacitor above self resonance. For most practical work however the native model will work fine because we should be using the inductor well below the self resonant frequency. A technique commonly used by microwave engineers to increase the Q factor of an inductor is to silver plate the wire. This can be modeled by setting Rho = 0.95 in the inductor model.

1.9 Chip Inductor Simulation in Genesys

The Genesys library has a collection of S parameter files for the Coilcraft chip inductors. Use the Part Selector to navigate to the Coilcraft chip inductors and select the 180 nH 1008 series inductor. The Part Selector and inductor schematic are shown in Figure 1-23.

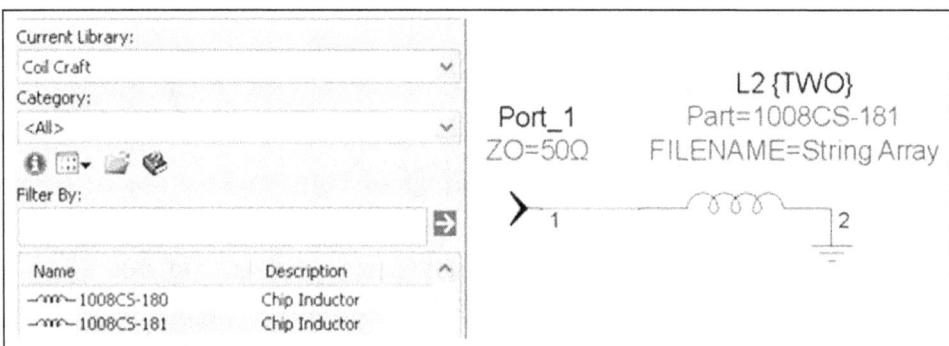

Figure 1-23 Part selector to locate chip inductor and the circuit schematic

Setup a Linear Analysis and plot the impedance of the inductor from 1 MHz to 1040 MHz.

Use Genesys Equation Editor to calculate and plot the inductance from the impedance, as shown in Figure 1-24. For reference the manufacturer's specifications for the inductance is overlaid on the plot.

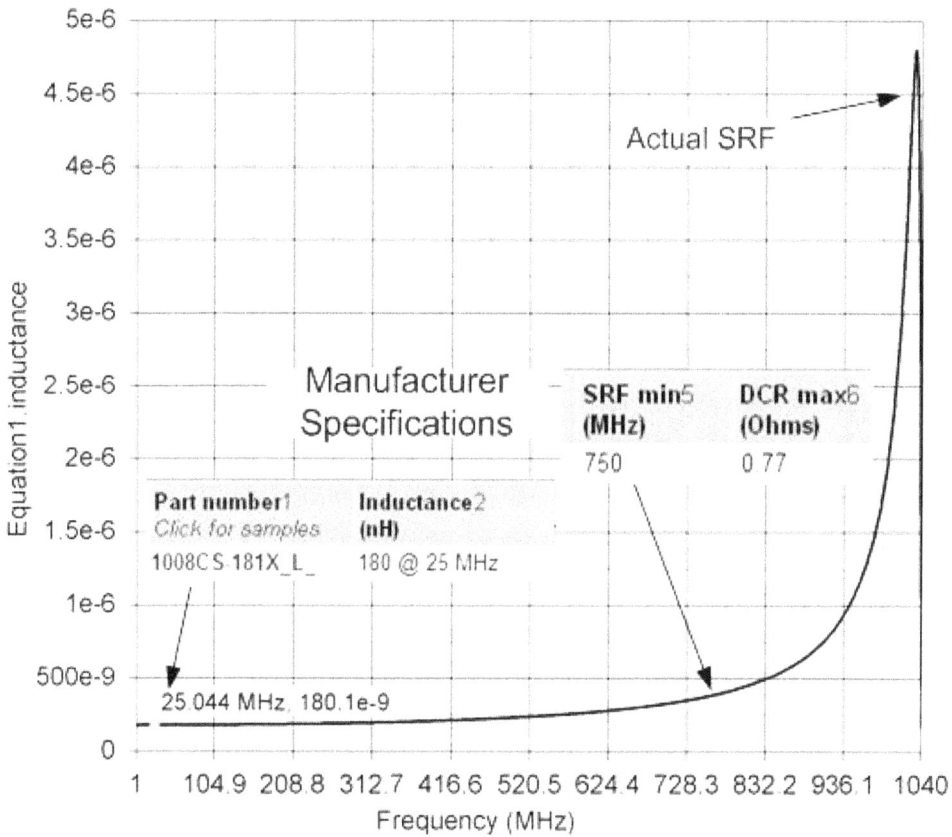

Figure 1-24 Plot of the 180 nH chip inductor inductance versus frequency

From the marker on the plot we see that the simulation of the S parameter file has very close correlation measuring 180.1 nH. The manufacturer specifies a minimum Q factor of 45 at a frequency of 100 MHz.

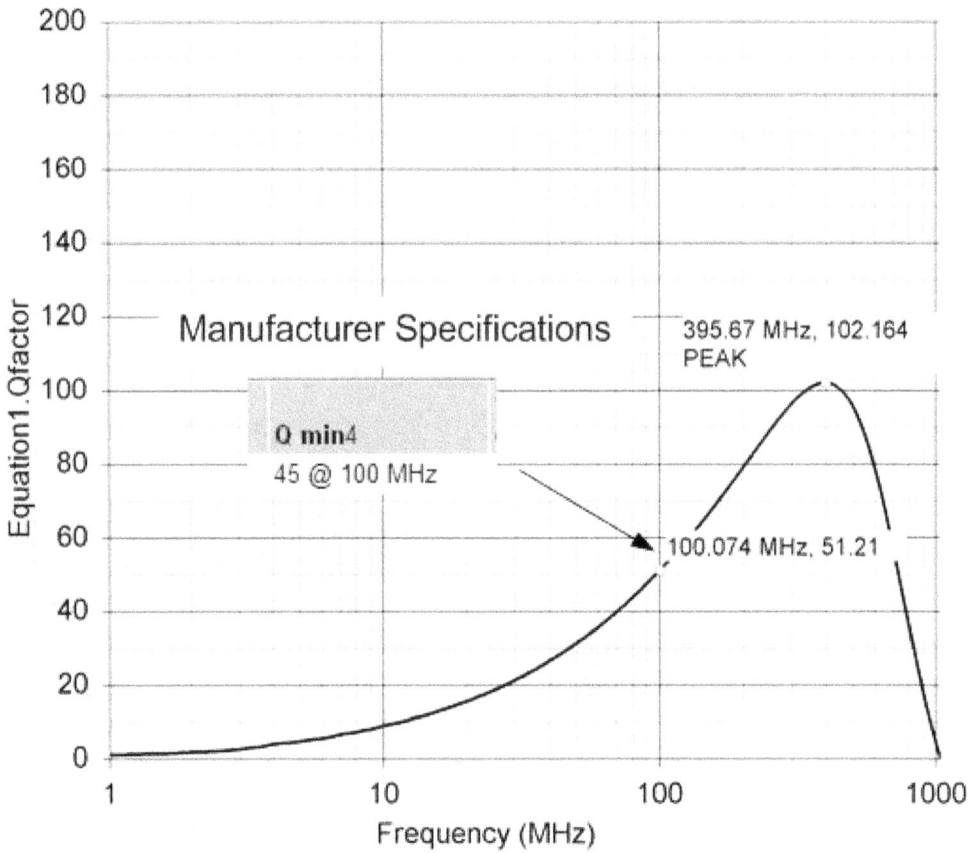

Figure 1-25 180 nH chip inductor Q factor versus frequency

The plot of the Q factor derived from the S parameter file shows that the Q factor is 51.21 near 100 MHz. Manufacturers will often plot the Q vs frequency on a logarithmic scale. It is very easy to change the x-axis to a logarithmic scale on the rectangular graph properties window in Genesys. This allows us to have a visual comparison to the manufacturer's catalog plot.

1.10 Magnetic Core Inductors

Design a 550 nH inductor using the Carbonyl W core of size T30. Determine the number of turns and model the inductor in Genesys.

From the manufacturer's data sheet the A_L value is 2.5 for a T30-10 toroidal core. Rearranging the equation to solve for the number of turns we find that 14.8 turns are required.

$$N = \sqrt{\frac{L}{A_L}} = \sqrt{\frac{550 \ nH}{2.5}} = 14.8$$

To reduce the winding loss we want to use the largest diameter of wire that will result in a single layer winding around the toroid. The following equation will give us the wire diameter.

$$d = \frac{\pi \ ID}{N + \pi}$$

Where:

d = Diameter of the wire in inches
ID = Inner diameter of the core in inches
N = Number of turns

Therefore,

$$d = \frac{\pi \ (0.151)}{14.8 + \pi} = \frac{0.4744}{19.942} = 0.0238 \ \ inches$$

From Appendix A, AWG#23 wire is the largest diameter wire that can be used to wind a single layer around the T30 toroid. Normally AWG#24 is chosen because this is a more readily available standard wire size. The toroidal inductor model in Genesys requires a few more pieces of information. The Genesys model requires that we enter the total winding resistance, core Q factor, and the frequency for the Q factor, F_q. As an approximation, set F_q to about six times the frequency of operation. In this case set F_q to (6)· 25 MHz = 150 MHz. Then tune the value of Q to get the best curve fit to the manufacturer's Q curve. We know that we have 14.8 turns on the toroid but we need to calculate the length of wire that these turns represent. The approximate wire length around one turn of the toroid is calculated from the following equation.

$$Length = \left[(2)\,Height + (OD - ID) \right] (\#turns)$$

Using the dimensions for the T30 toroid from Table 1-1 we can calculate the total length of the wire as 6.10 inches.

$$(2)(0.128) + (0.307 - 0.151)] \cdot 14.8 = 6.10\ inches$$

Then use the techniques covered previously to calculate the resistance of the 6.10 inches of wire taking into account the skin effect. To get a better estimate of the actual inductor Q, use the F_q frequency rather than the operating frequency for the skin effect calculation. The resistance at 150 MHz for the AWG#24 wire is calculated as 0.30 Ω. Use an Equation Editor to calculate the Q and inductance. Figure 1-26 shows the schematic of the toroidal inductor and the simulated Q.

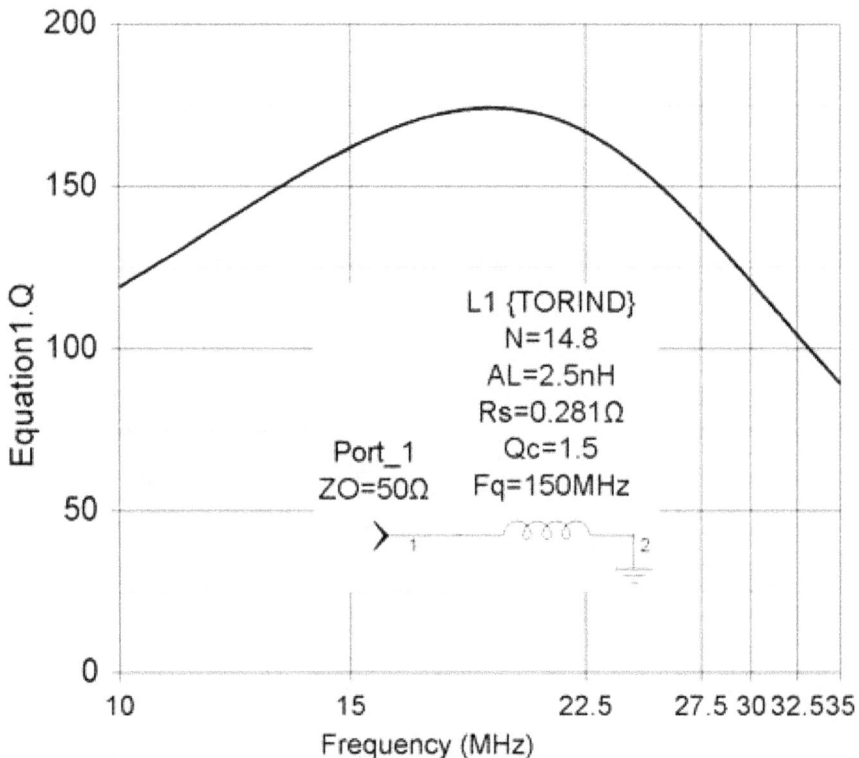

Figure 1-26 Genesys toroidal inductor model and simulated Q factor

The Qc parameter of the model has been tuned to give a reasonably good fit with the manufacturer's (T30) curve. Figure 1-27 shows that the simulated self resonant frequency of the inductor model is near 150 MHz.

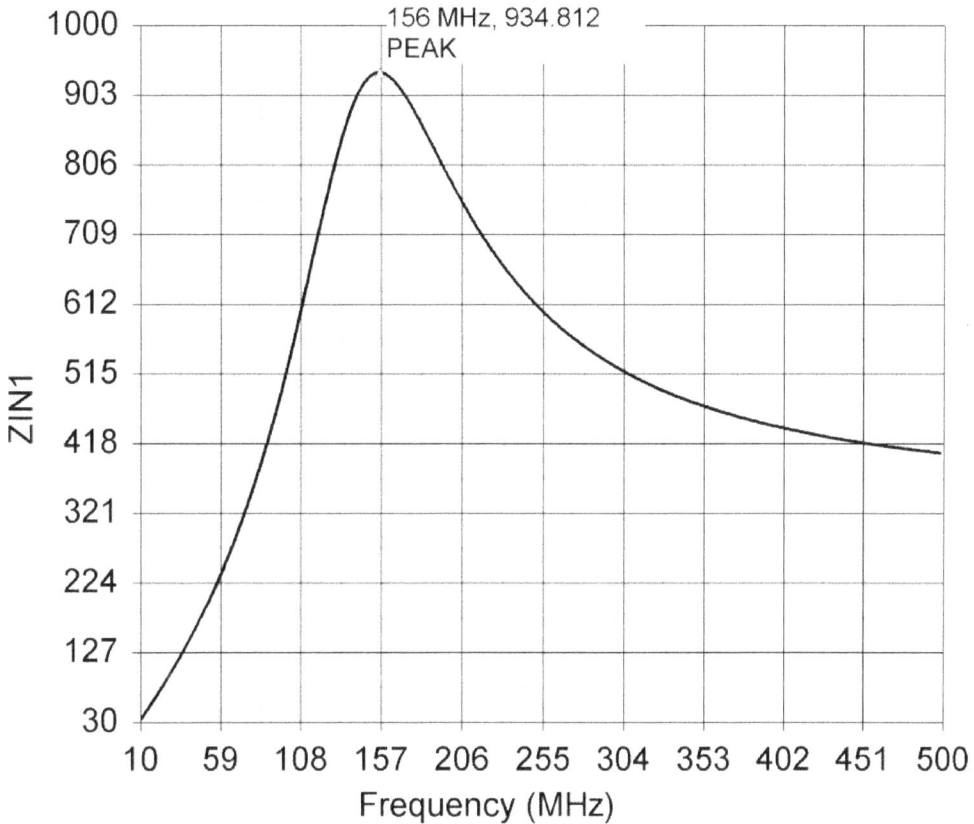

Figure 1-27 Impedance of toroidal inductor model

Figure 1-28 shows the inductance that was calculated from the impedance. The inductance is exactly 550 nH at 10 MHz and begins to increase slightly to 563 nH at the design frequency of 25 MHz. Given the myriad of variables associated with the toroidal inductor, this simple model gives a good first order model of the actual inductor and will enable accurate simulation of filter or resonator circuits.

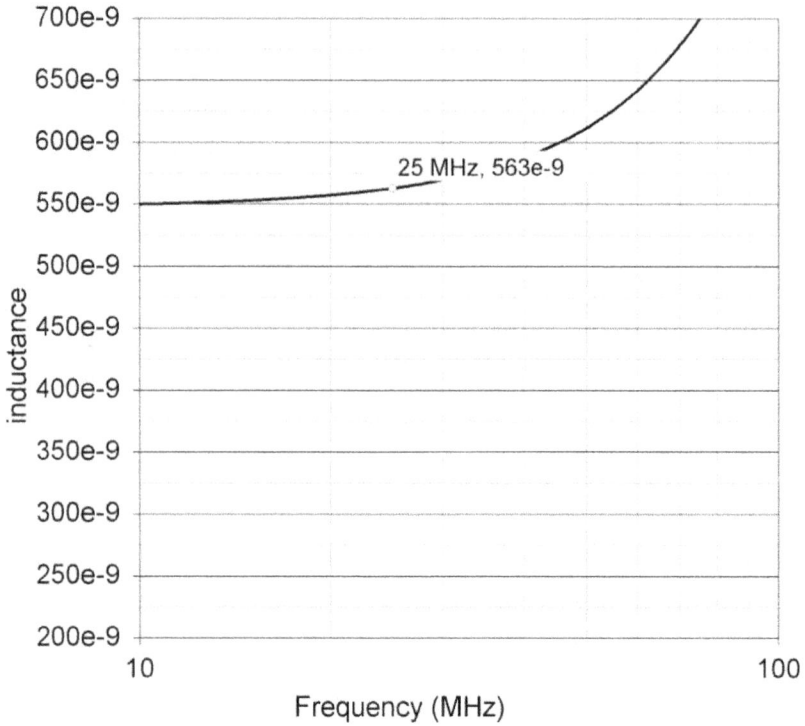

Figure 1-28 Inductance value of the toroidal inductor model

1.11 Single Layer Capacitor

The single layer capacitor is one of the simplest and most versatile of the surface mount capacitors. It is formed with two plates that are separated by a single dielectric layer as shown in Figure 1-29. Most of the electric field (E) is contained within the dielectric however there is a fraction of the E field that exists outside of the plates. This is known as the fringing field.

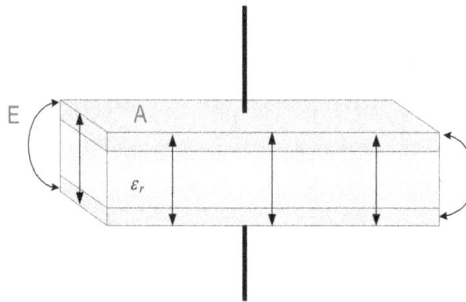

Figure 1-29 Single layer parallel plate capacitor

The capacitance formed by a dielectric material between two parallel plate conductors is given by the following equation.

$$C = (N-1) \left(\frac{KA\varepsilon_{\mathrm{r}}}{t} \right) (FF) \quad pF$$

Where,

A = plate area
ε_r = relative dielectric constant
t = separation
K = unit conversion factor; 0.885 for cm and 0.225 for inches
FF = fringing factor; 1.2 when mounted on microstrip
N = number of parallel plates.

Example 1-10: Design a single layer capacitor from a dielectric that is 0.010 inches thick and has a dielectric constant of three. Each plate is cut to 0.040 inches square.

Solution: When the capacitor is mounted with at least one plate on a large printed circuit board track, a value of 1.2 is typically used in calculation. The MATLAB equations can be used to solve the capacitor value as shown in Figure 1-30.

```
er=3
t=0.010
A=0.0016
N=2
FF=1.2
K=0.225
C=(N-1)*FF*((K*A*er)/t)
```

Figure 1-30 Single layer capacitance calculation

The single layer capacitor can be modeled in Genesys using the Thin Film Capacitor.

$$C = (2-1) \left(\frac{(.225)(0.04 \cdot 0.04)(3)}{0.010} \right) (1.2) = 0.13 \, pF$$

The ceramic dielectrics used in capacitors are divided into two major classifications. Class 1 dielectrics have the most stable characteristics in terms of temperature stability. Class 2 dielectrics use higher dielectric constants which result in higher capacitance values but have greater variation over temperature. The temperature coefficient is specified in either percentage of nominal value or parts per million per degree Celsius (ppm/^{o}C). Ceramic materials with a high dielectric constant tend to dominate RF applications with a few exceptions. NPO (negative-positive-zero) is a popular ceramic that has extremely good stability of the nominal capacitance versus temperature.

Dielectric Material	Dielectric Constant
Vacuum	1.0
Air	1.004
Mylar	3
Paper	4 - 6
Mica	4 - 8
Glass	3.7 - 19
Alumina	9.9
Ceramic (low ε_r)	10
Ceramic (high ε_r)	100 – 10,000

Table 1-2 Dielectric constants of materials

1.12 Capacitor Physical Model and Q Factor

RF losses in the dielectric material of a capacitor are characterized by the dissipation factor. The dissipation factor is also referred to as the loss tangent and is the ratio of energy dissipated to the energy stored over a period of time. It is essentially the capacitor's efficiency rating. The dissipation factor and other ohmic losses lead to a parameter known as the Equivalent Series Resistance, ESR. The dissipation factor is the reciprocal of the Q factor. Just as we have seen with resistors and inductors, the physical model of a capacitor is a network of R, L, and C components.

Example 1-11: Calculate the Q factor versus frequency for the physical model of an 8.2 pF multilayer chip capacitor shown in Figure 1-31.

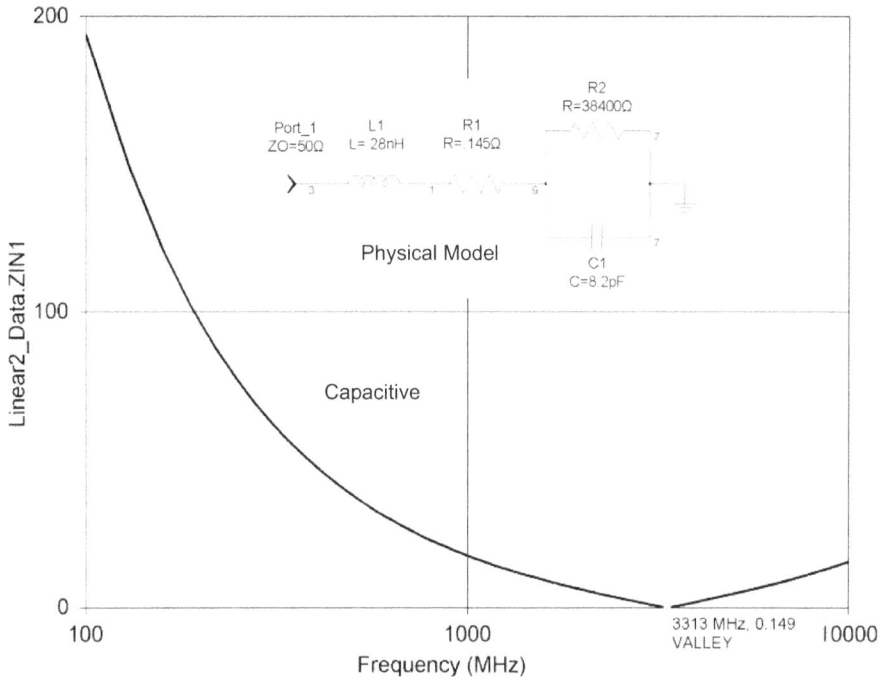

Figure 1-31 Physical model of the 8.2 pF chip capacitor and impedance

The capacitor has a series inductance and resistance component along with a resistance in parallel with the capacitance. The parallel resistor sets the losses in the dielectric material. The series resistance and inductance represent any residual lead inductance and ohmic resistances.

Solution: The values entered for the physical model and the Q factor can be obtained from the capacitor manufacturer. The plot of Figure 1-31 shows the impedance of the capacitor versus frequency. Note that the impedance decreases as would be expected until the self resonant frequency is reached. Above the self resonant frequency the impedance begins to increase suggesting that the capacitor is behaving as an inductor. The self resonant frequency is due to the series inductance resonating with the capacitance. At resonance the reactance cancels leaving only the resistances *R1* and *R2*. The parallel resistance, *R2*, can be converted to an equivalent series resistance by the following equation.

$$R2' = \frac{1}{1+Q^2}\ R2$$

These two series resistances can then be added to find the equivalent series resistance, ESR:

$$ESR = R1 + R2'$$

The capacitor Q factor is then calculated by the next equation. Note that X_T is the total series reactance of the inductive and capacitive reactance.

$$Q = \frac{X_T}{ESR}$$

As Figure 1-31 shows, the 8.2 pF multilayer chip capacitor has a series resonant frequency of 3313 MHz. A marker is placed on the trace indicating the impedance at the SRF as 0.149 Ω. Because the reactance is cancelled at the SRF, this impedance essentially becomes the ESR of the capacitor. R2 of Figure 1-31 is extremely frequency dependent. This means that the capacitor's Q factor is also extremely frequency dependent. An improved model for analyzing the capacitor's characteristics over a wide frequency range is to use the Genesys model for capacitor with Q. Using this model the Q factor of the capacitor can be made proportional to the square root of the applied frequency. The capacitor Q factor can be calculated from the impedance using the Equation Editor as shown in Figure 1-32. The resistive and inductive components are used to define the Q factor.

```
using("Linear1_Data")
reactance1=im(Linear1_Data.ZIN1)
resistance1=re(Linear1_Data.ZIN1)
Qphysical=abs(reactance1)./resistance1
capacitance_model=1/(abs(reactance1).*F*2*3.14)
```

Figure 1-32 MATLAB equations used to calculate Q factor and capacitance

Figure 1-33 shows the large dependence on the capacitor Q with frequency. At 1006 MHz the Q factor is 120 while at 100 MHz the Q factor is greater than 1000. The Q factor goes to zero at the self resonant frequency. Above the self resonant frequency the Q is undefined.

Example 1-12: Calculate the Q factor versus frequency for the modified physical model of an 8.2 pF multilayer chip capacitor shown in Figure 1-33.

Solution: The modified model is shown in Figure 1-33.

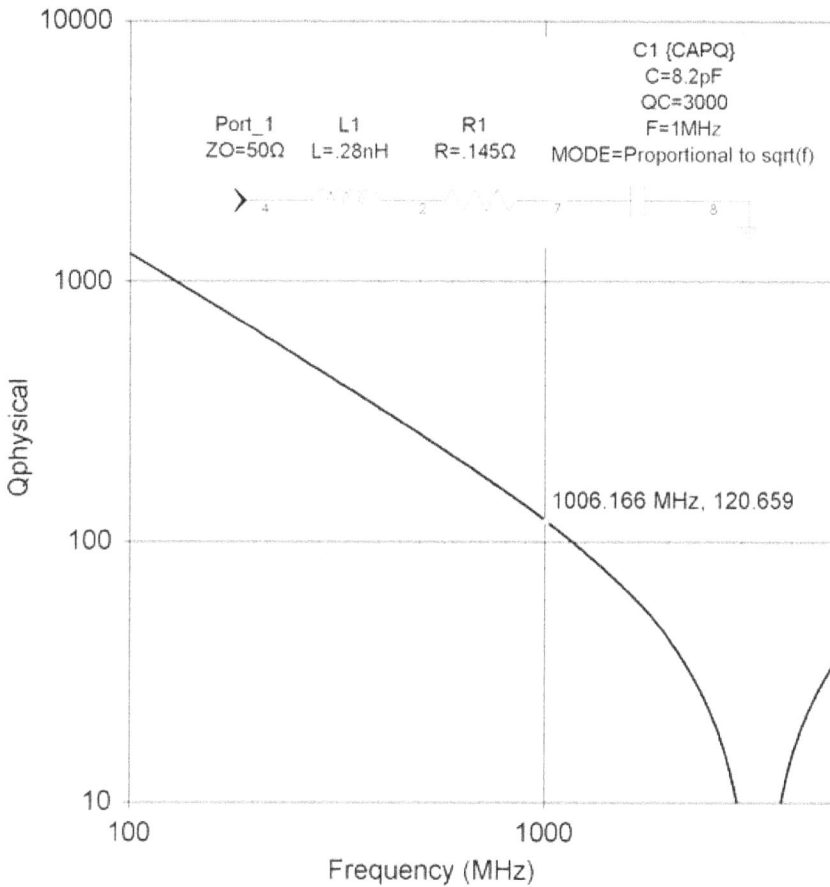

Figure 1-33 Q factor plot of 8.2 pF chip capacitor and equivalent model

The MATLAB equation of Figure 1-32 calculates the effective capacitance from the total reactance of the model using the following equation.

$$C=\frac{1}{2\pi F X_T}$$

The plot of Figure 1-34 reveals some interesting characteristics about the chip capacitor. From 100 MHz to 300 MHz the capacitance value is fairly constant at 8.217 pF. As the frequency increases above 300 MHz we see that the capacitance actually increases. The parasitic inductive reactance of the capacitor package actually makes the effective capacitance greater than its nominal value.

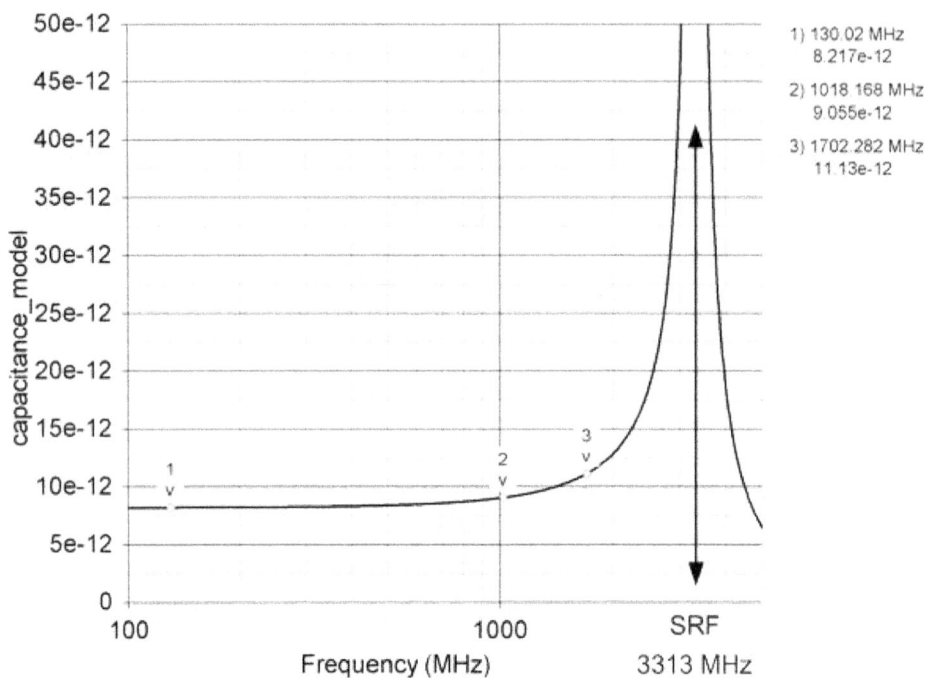

Figure 1-34 Effective capacitance of the 8.2 pF chip capacitor

This is a property of the capacitor that is not always intuitive. As the frequency approaches the self resonant frequency the capacitance rapidly approaches infinity. The capacitor actually becomes nearly a short circuit to RF at the self resonant frequency. This is an important property of the capacitor that is used frequently in RF and microwave design. In RF coupling or bypass capacitor applications, capacitors are very often used at or near their self resonant frequency. The capacitor's self resonant frequency is due to the series resonant circuit. In applications requiring bypassing over a wide range of frequencies it is often necessary to use several capacitors each selected to have a uniquely spaced self resonant frequency. In filter and other tuned circuit applications where we want the chip to appear as an 8.2 pF capacitor, we clearly must stay well below the self resonant frequency of the capacitor. A typical rule-of-thumb is to use the capacitor over a frequency range up to 35% of the self resonant frequency. Therefore the 8.2 pF chip capacitor with a self resonant frequency of 3313 MHz would be used as a capacitor in tuned circuits up to a frequency of about 1160 MHz.

References and Further Reading

[1]   Ali A. Behagi and Stephen D. Turner, *Microwave and RF Engineering,* A Simulation Approach with Keysight Genesys Software, BT Microwave LLC, March, 2015

[2]   Paul Lorrain, Dale P. Corson, and Francois Lorrain, *Electromagnetic Fields and Waves*, W.H. Freeman and Company, New York, 1988

[3]   *Design Guide, Microwave Components* Inc., P.O. Box 4132, South Chelmsford, MA 01824

[4]   *Iron Powder Cores for High Q Inductors*, Micrometals, Inc.

[5]   Keysight Technologies, *Genesys 2015.08, Users Guide*, www.keysight.com

[6]   *Capacitors for RF Applications,* Dielectric Laboratories, Inc.,2777 Rt.20 East, Cazenovia, NY. 13035

[7]   R. Ludwig and P. Bretchko, *RF Circuit Design -Theory and Applications*, Prentice Hall, New Jersey, 2000

[8]   William Sinnema and Robert McPherson, *Electronic Communication,* Prentice Hall Canada, 1991

[9]   M.F. "Doug" DeMaw, *Ferromagnetic Core Design & Application Handbook*, MFJ Publishing Co., Inc. Starkville, MS. 39759, 1996

[10]   *The RF Capacitor Handbook,* American Technical Ceramics, One Norden Lane, Huntington, New York 11746

Problems

1-1.   Calculate the wavelength of an electromagnetic wave operating at a frequency of 428MHz.

1-2.   Calculate the inductance of a 5 inch length of AWG #30 straight copper wire.

1-3.   Using the Equation Editor, calculate the reactance of the wire from Problem 2 at 10 Hz, 10 MHz, and 10 GHz. Create a Linear analysis in Genesys and display the wire impedance vs. frequency.

1-4.   Calculate the resistance of a 12 inch length of AWG #24 copper wire at DC and at 25MHz

1-5.   Find the skin depth and the resistance of a 2 meter length of copper coaxial line at 2 GHz. The inner conductor radius is 1 mm and the outer conductor is 4 mm.

1-6.   Calculate the inductance of a 5 inch length of copper flat ribbon conductor. The dimensions of the ribbon are 0.100 inches in width and 0.002 inches thick.

1-7.   Model a chip resistor (size 0603) with a resistance of 50Ω in Genesys. Consider an application in which 50Ω impedance must be maintained with +10%. Create a Linear Analysis and determine the maximum usable frequency of the chip resistor.

1-8.   Design an air core inductor with an inductance value of 84nH. Use a copper wire of 0.050 inch diameter wound on a core diameter or 0.100 inch. Determine the number of turns required assuming a tight spaced winding.

1-9.   Using the inductor from Problem 8 and the techniques developed in Section 1.4.1 examine the change in the coil inductance as the turn spacing is increased from zero to 0.10 inch in 0.002 inch increments.

1-10.   Using the inductor from Problem 8, determine the self resonant frequency of the inductor and comment on the maximum frequency in which the inductor may be used in a tuned circuit application.

1-11.   Using the inductor from Problem 8, determine the maximum Q factor of the inductor and the frequency at which the maximum Q factor is obtained.

1-12.   In Genesys, select a chip inductor from the CoilCraft library with an inductance value of 80nH. Determine the maximum Q factor of the inductor and comment on the maximum usable frequency of the inductor in a filter application.

1-13.   Design a 1mH toroidal inductor on a Carbonyl W core size T30. Determine the maximum wire size that could be used to realize a single layer winding.

1-14.   Using the inductor from Problem 13, model the inductor in Genesys and determine the approximate self resonant frequency. Comment on the maximum usable frequency of the inductor in the front end of a radio receiver.

1-15.   A 0.05pF capacitor is required to couple a transistor to the resonator of a microwave oscillator. Design a single layer capacitor using a 0.020 inch thick dielectric with e_r=2.2. Determine the dimensions of the capacitor assuming square footprint is desired.

1-16.   For the single layer capacitor of Problem 15, determine the dimensions of the capacitor with a dielectric constant e_r=10.2.

1-17.   In Genesys, model a 47pF chip capacitor using the ATC 0603 model from the Modelithics library. Determine the Q factor of the capacitor at a frequency of 1000MHz.

1-18.   Determine the self resonant frequency of the capacitor in Problem 17 and comment on the maximum frequency that this capacitor could be used in a tuned circuit. At what frequency does the capacitor have the lowest amount of energy loss?

Chapter 2

Transmission Lines

2.1 Introduction

Transmission lines play an important role in designing RF and microwave networks. In chapter 1 we have seen that, at high frequencies where the wavelength of the signal is smaller than the dimension of the components, even a small piece of wire acts as an inductor and affects the performance of the network.

Example 2-1: For the series RLC elements in Figure 2-1 measure the reflection coefficients and VSWR from 100 to 1000 MHz in 100 MHz steps.

Solution: Create a Linear Analysis in Genesys and sweep the frequency from 100 to 1000 MHz to measure the VSWR at port 1 and the input reflection coefficients, S[1,1], in dB and (Mag abs+Angle) formats.

Port_1 ZO=50Ω X1 R=55Ω L=15nH C=20pF	F (MHz)	VSWR1	S[1,1] (dB)	mag(S[1,1])	ang(S[1,1]) (deg)
	100	3.514	-5.084	0.557	-52.175
	200	1.503	-13.933	0.201	-65.292
	300	1.106	-25.944	0.05	18.321
	400	1.42	-15.207	0.174	64.69
	500	1.811	-10.796	0.289	64.345
	600	2.245	-8.321	0.384	61.007
	700	2.727	-6.682	0.463	57.291
	800	3.26	-5.506	0.531	53.694
	900	3.849	-4.62	0.588	50.344
	1000	4.494	-3.931	0.636	47.27

Figure 2-1 Table of VSWR, return loss, and reflection coefficient

2.2 Return Loss, VSWR, and Reflection Coefficient Conversion

Return Loss, VSWR, and Reflection Coefficient are all different ways of characterizing the wave reflection. These definitions are often used interchangeably in practice.

Example 2-2: Generate a table showing the return loss, the reflection coefficient, and the percentage of reflected power as a function of VSWR.

Solution: In Genesys create a schematic with a resistor. Make the resistance value a tunable variable. Set the Linear Analysis at a fixed frequency of 100 MHz. Then add a Parameter Sweep to sweep the value of the resistor.

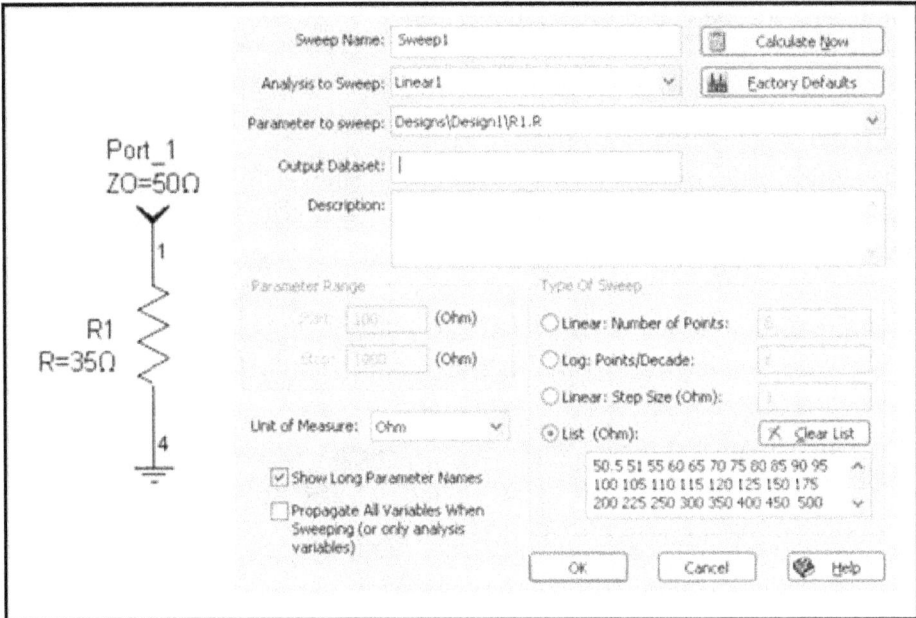

Figure 2-2 Schematic and parameter sweep for VSWR table

Edit the Parameter Sweep and select the current Linear Analysis. Then select the resistance of the Resistor element on the Parameter to Sweep drop down list. Under the Type of Sweep choose the List option and enter the discrete resistance values as shown in Figure 2-2. When the circuit is swept the 30 resistance values will create a unique VSWR, Return Loss, and Reflection Coefficient. Then create an equation to calculate this parameter. Figure 2-3 shows the equation block for the calculation of mismatch loss.

```
vswr=Sweep1_Data.VSWR1
Reflcoef=(vswr-1)/(vswr+1)
mismatch=1-((ReflCoef)^2)
powerloss=(1-mismatch)*(100)
```

Figure 2-3 MATLAB equation block for calculation of mismatch loss

We will present the mismatch loss as a percentage of the available power that is reflected by the load. Add the equation block variable, power loss, to the output table. The results can be read into an attractive attractive table shown in Table 2-1. As we can see from the table if we can keep the VSWR less than 1.25:1 we will have less than 1% power loss due to reflective impedance mismatch.

VSWR	Return Loss (dB)	$\lvert\Gamma\rvert$	% Reflected Power	VSWR	Return Loss (dB)	$\lvert\Gamma\rvert$	% Reflected Power
1.01	-46.06	0.00	0.00%	2.40	-7.71	0.41	16.96%
1.02	-40.09	0.01	0.01%	2.50	-7.36	0.43	18.37%
1.10	-26.44	0.05	0.23%	3.00	-6.02	0.50	25.00%
1.20	-20.83	0.09	0.83%	3.50	-5.11	0.56	30.86%
1.30	-17.69	0.13	1.70%	4.00	-4.44	0.60	36.00%
1.40	-15.56	0.17	2.78%	4.50	-3.93	0.64	40.50%
1.50	-13.98	0.20	4.00%	5.00	-3.52	0.67	44.44%
1.60	-12.74	0.23	5.33%	6.00	-2.92	0.71	51.02%
1.70	-11.73	0.26	6.72%	7.00	-2.50	0.75	56.25%
1.80	-10.88	0.29	8.16%	8.00	-2.18	0.78	60.49%
1.90	-10.16	0.31	9.63%	9.00	-1.94	0.80	64.00%
2.00	-9.54	0.33	11.11%	10.00	-1.74	0.82	66.94%
2.10	-9.00	0.36	12.59%	20.00	-0.87	0.91	81.86%
2.20	-8.52	0.38	14.06%	200.00	-0.09	0.99	98.02%
2.30	-8.09	0.39	15.52%	2000.00	-0.01	1.00	99.80%

Table 2-1 Relationship among return loss, VSWR, and reflection coefficient

2.4 Waveguide Transmission Lines in Genesys

Because of the complex EM fields that can propagate in a waveguide, modern computer aided design techniques are best handled by three dimensional EM solvers. Genesys has two models of the waveguides that are useful to engineers. The first model is a straight section of the waveguide in which the $TE_{1,0}$ mode is utilized. The second model is a waveguide to TEM transition which is similar to an adapter. The input and output ports used in Genesys can be thought of as coaxial ports supporting TEM propagation. Therefore we cannot attach a section of waveguide directly to a port because an impedance mismatch will exist.

Example 2-3: Measure and display the insertion loss of a three inch length of WR 112 waveguide from 4 to 8 GHz.

Solution: Make the length of the waveguide tunable. From Table 2-2 we can get the width (a) and height (b) dimensions to enter into the waveguide model.

Frequency Band, GHz	U.S. (EIA) Designator	British WG Designator	Cut Off Freq. in GHz $TE_{1,0}$	a dimension inches	b dimension inches
1.12 - 1.70	WR 650	WG 6	0.908	6.500	3.250
1.45 - 2.20	WR 510	WG 7	1.158	5.100	2.550
1.70 - 2.60	WR 430	WG 8	1.375	4.300	2.150
2.20 - 3.30	WR 340	WG 9A	1.737	3.400	1.700
2.60 - 3.95	WR 284	WG 10	2.080	2.840	1.340
3.30 - 4.90	WR 229	WG 11A	2.579	2.290	1.145
3.95 - 5.85	WR 187	WG 12	3.155	1.872	0.872
4.90 - 7.05	WR 159	WG 13	3.714	1.590	0.795
5.85 - 8.20	WR 137	WG 14	4.285	1.372	0.622
7.05 - 10.00	WR 112	WG 15	5.260	1.122	0.497
8.2 - 12.4	WR 90	WG 16	6.560	0.900	0.400
9.84 - 15.0	WR 75	WG 17	7.873	0.750	0.375
11.9 - 18.0	WR 62	WG 18	9.490	0.622	0.311
14.5 - 22.0	WR 51	WG 19	11.578	0.510	0.255
17.6 - 26.7	WR 42	WG 20	14.080	0.420	0.170
21.7 - 33.0	WR 34	WG 21	17.368	0.340	0.170
26.4 - 40.0	WR 28	WG 22	21.100	0.280	0.140
32.9 - 50.1	WR 22	WG 23	26.350	0.224	0.112
39.2 - 59.6	WR 19	WG 24	31.410	0.188	0.094
49.8 - 75.8	WR 15	WG 25	39.900	0.148	0.074
60.5 - 91.9	WR 12	WG 26	48.400	0.122	0.061
73.8 - 112.0	WR 10	WG 27	59.050	0.100	0.050

Table 2-2 Standard rectangular waveguide characteristics

Set the source and load resistors equal to 377 Ω (simulating the waveguide to TEM adapter) representing the impedance of free space. Sweep the insertion loss (S21) from 4 GHz to 8 GHz as shown in Figures 2-4 and 2-5.

Note that the use of the waveguide models does require a substrate definition. In this case however the waveguide models only use the dielectric constant Er, and Rho, the resistivity of the metal walls. Typically enter values of one for both the air dielectric and the resistivity normalized to copper. Figure 2-3 shows the schematic of the waveguide. The insertion loss of the waveguide in its pass band at 8 GHz is extremely low. This is one of the advantages of using waveguide transmission lines as they are practically the lowest loss microwave transmission line available. Also note that the insertion loss increases as we move below the cutoff frequency.

A marker is placed at the cutoff frequency of 5260 MHz. Increase the length of the waveguide to 3 inches. Note the dramatic increase in the rejection below the cutoff frequency. The insertion loss in band is still quite low. We can see that using the waveguide below the cutoff frequency is an effective method of achieving a very good microwave high pass filter.

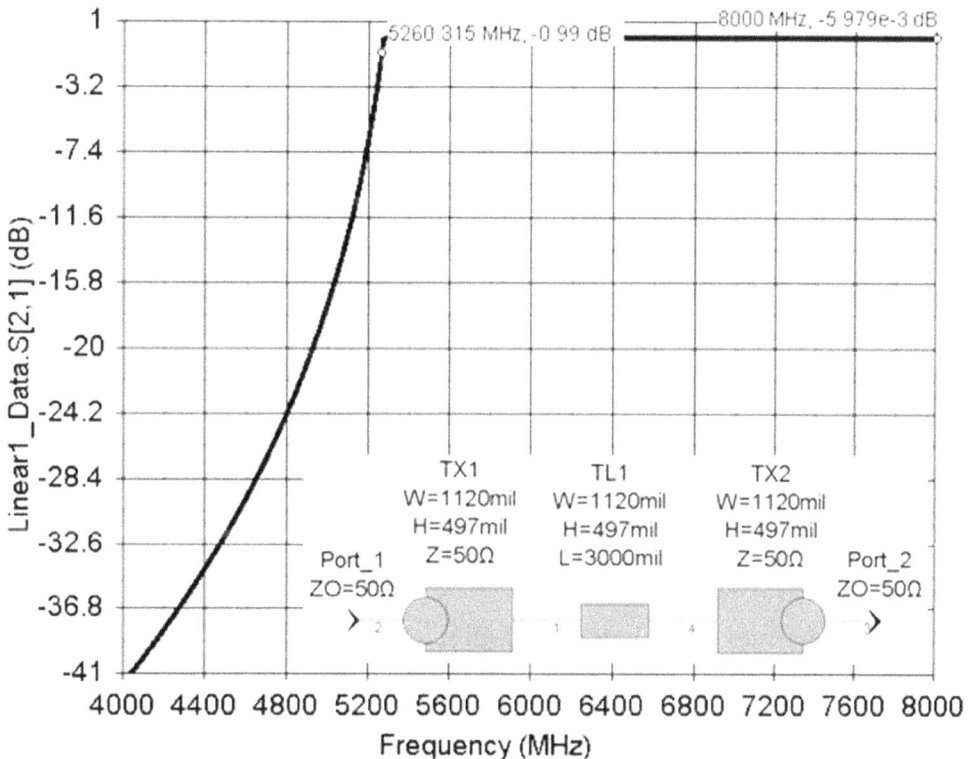

Figure 2-4 Three inch WR112 waveguide with TEM adapters

2.4 Group Delay in Transmission Lines

A frequently encountered concept related to the transmission line velocity factor is group delay. Group delay is a measure of the time that it takes a signal to traverse a transmission line, or its transit time. It is a strong function of the length of the line, and usually a weak function of frequency. It is expressed in units of time, picoseconds for short distances or nanoseconds for longer distances. Remember that in free space all electromagnetic signals travel at the speed of light, c, which is approximately 300,000 kilometers per second. Therefore, in free space, electromagnetic radiation travels one foot in one nanosecond, unless there is something to slow it down such as a dielectric. Mathematically the group delay is the derivative of phase versus frequency. In communication systems, the ripple in the group delay creates a form of distortion.

2.5 Comparing Group Delays of Transmission lines

Example 2-4: To compare group delays of various transmission lines, create a Genesys workspace with four schematics and corresponding linear analysis. In each schematic model place a 20 inch length of the previously discussed transmission lines. Use the RG8 cable for the coaxial transmission line. For the microstrip and stripline transmission lines use the Rogers RO3003 dielectric material with 1 oz. copper and 30 mil substrate thicknesses.

Solution: The TLINE utility can be used to synthesize the line widths for 50 Ω lines. Use WR430 for the waveguide transmission line. Refer to Table 2-3 for the waveguide dimensions. The schematics representing the four transmission lines are shown in Figure 2-5. Sweep the frequency from 2 GHz to 5 GHz in each of the four transmission lines. Genesys has a built-in function for the display of the group delay. The group delay (gd) function is shown in the inset of Figure 2-6.

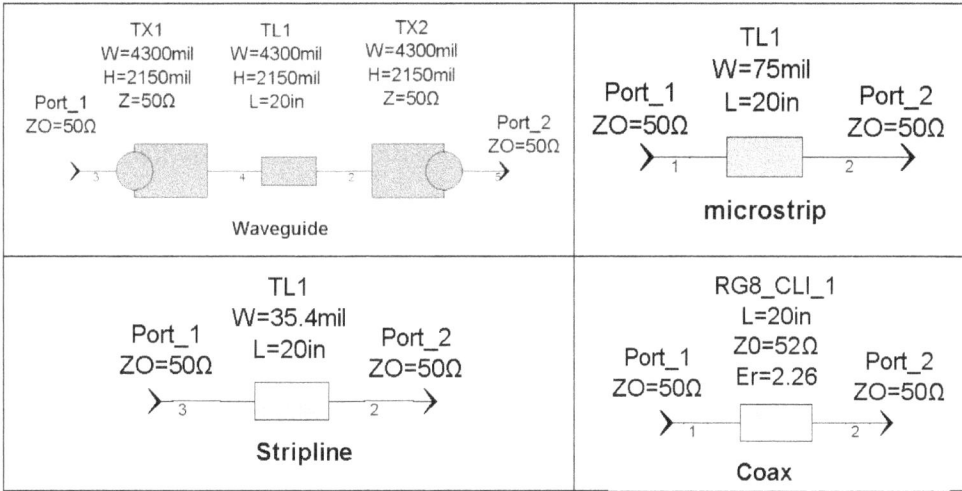

Figure 2-5 20 inch length of different transmission lines

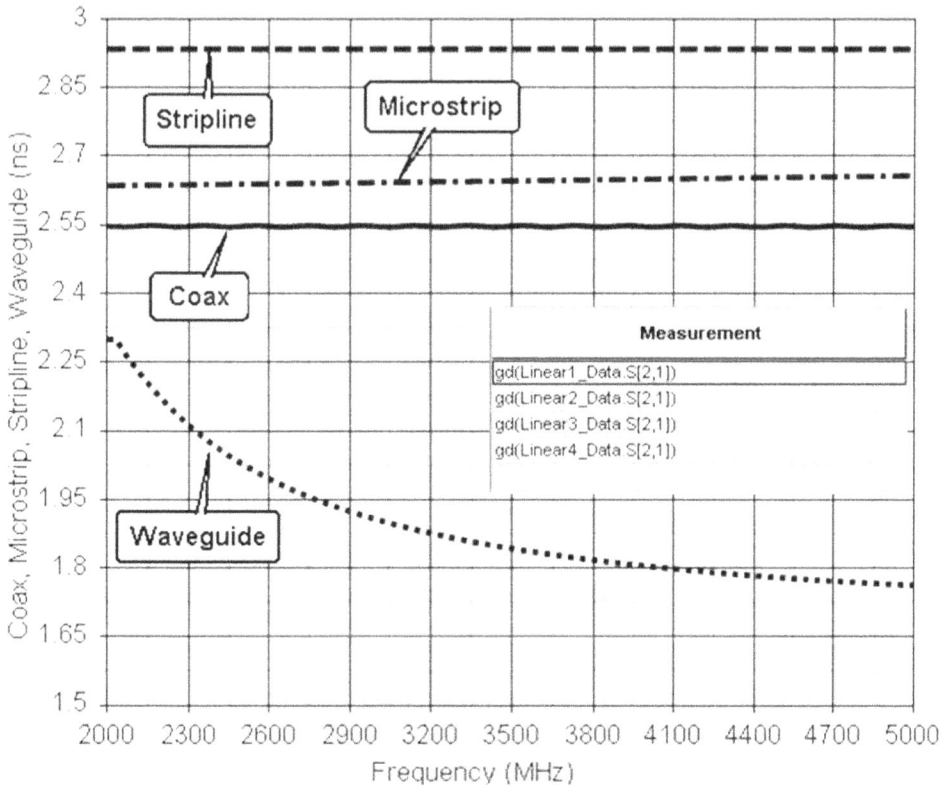

Figure 2-6 Transmission line group delay comparison

2.6 Short-Circuited Transmission Lines

It has been demonstrated that the input impedance of a lossless short-circuited transmission line is a pure imaginary function; and the input reactance is given by the following equation.

$$X_{in} = Z_O \tan \theta$$

where $\beta = \theta d$ is the electrical length of the transmission line in degrees

We see in Figure 2-7 that this reactance can change from inductive to capacitive depending on the length of the transmission line.

Example 2-5: Plot the reactance of a loss less short-circuited transmission line as a function of the electrical length of the line.

Solution: To plot the reactance of a lossless short-circuited transmission line in Genesys, create a schematic with a grounded transmission line. Make the length of the transmission line a variable with any starting value in degrees. Setup a Linear Analysis with a single frequency at 1500 MHz. Then use a Parameter sweep to vary the electrical length of the transmission line from 0 to 360 degrees. Setup a graph to plot the reactance of the shorted transmission line vs. the electrical length from the Parameter sweep data set as shown in Fig. 2-7.

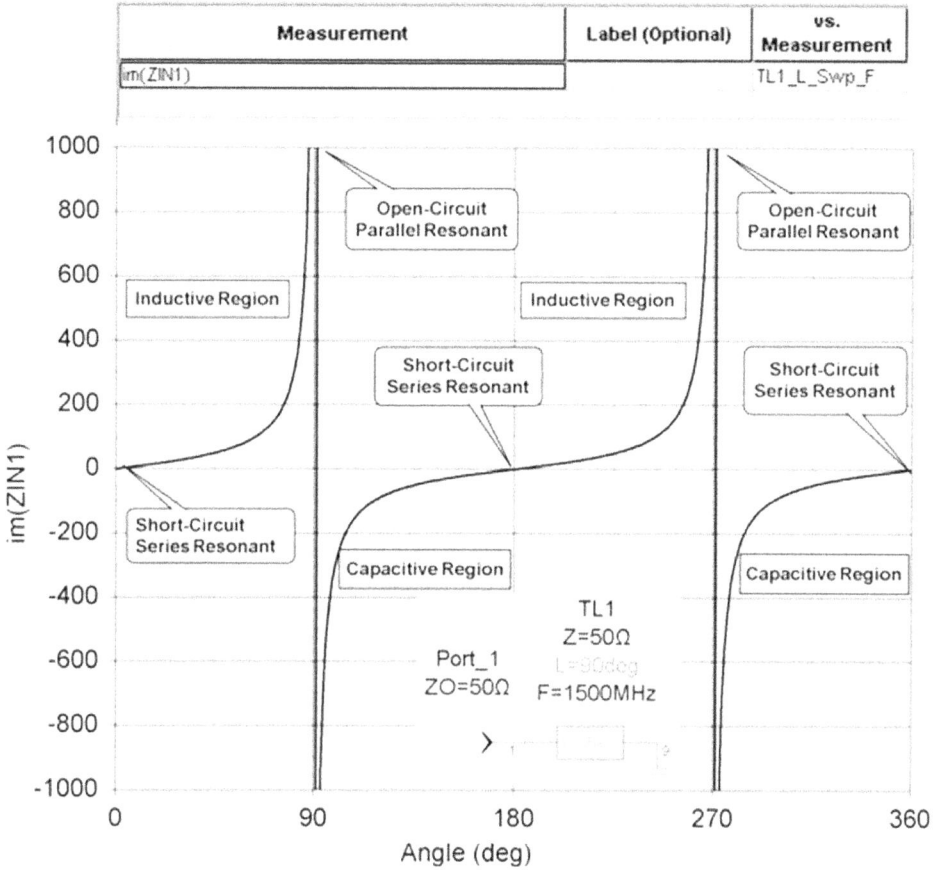

Figure 2-7 Short-circuited line reactance versus electrical length

Example 2-6: Calculate the physical length of a short circuited microstrip transmission line for a given electrical length of the line.

Solution: In Genesys simply place an ideal transmission line element on a schematic and enter the desired impedance, electrical length, and frequency. Select the Advanced TLINE utility from the Schematic menu. The microstrip line will be created as shown in Figure 2-8.

Ideal Transmission Line
TL1
Z=50Ω
Port_1   L=90deg
ZO=50Ω   F=1500MHz

Physical Microstrip Line
TL1
Port_1   W=73.617mil
ZO=50Ω   L=1268.764mil

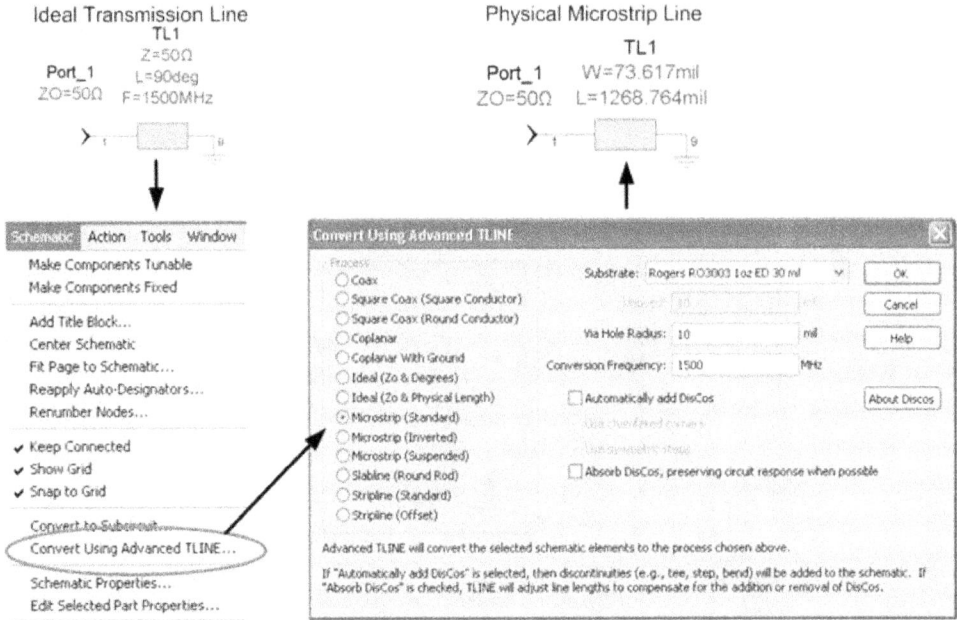

Figure 2-8 Advanced TLINE for calculation of microstrip line length

Figure 2-9 shows the correct method of modeling a microstrip short-circuited transmission line with a via hole. As the table inset of Figure 2-9 shows, the impedance of a shorted quarter-wave section is close to an open circuit. This type of line section could be used as a parallel resonant circuit.

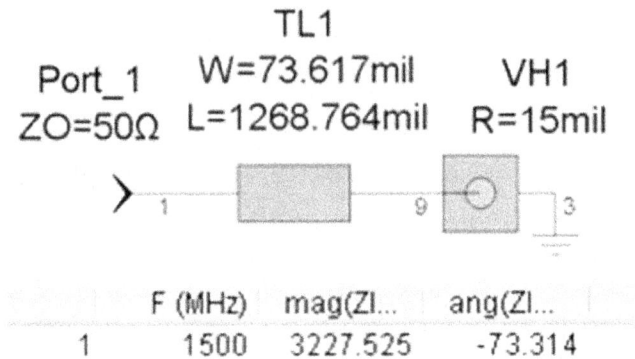

TL1
Port_1   W=73.617mil   VH1
ZO=50Ω   L=1268.764mil   R=15mil

F (MHz)	mag(Zl...	ang(Zl...
1	1500	3227.525	-73.314

Figure 2-9 Quarter wave short-circuited line

2.7 Open-Circuited Transmission Lines

It has been shown that the input impedance of a lossless open circuited transmission line is a pure imaginary function; and the input reactance is given by the following equation.

$$X_{in} = Z_O \cot \theta$$

where: θ is the electrical length of the transmission line in degrees.

Following the same procedures, use a parameter sweep in Genesys to observe the behavior of this reactance as the length of the open circuit transmission line is varied from 0 to 360 degrees. Note that the transmission line is terminated with a 10^6 Ω load to emulate an open circuit termination on the transmission line in Figure 2-10. Comparing the open circuit reactance to the short-circuited line reactance we can see that a 90°, $\lambda_g/4$, offset is present.

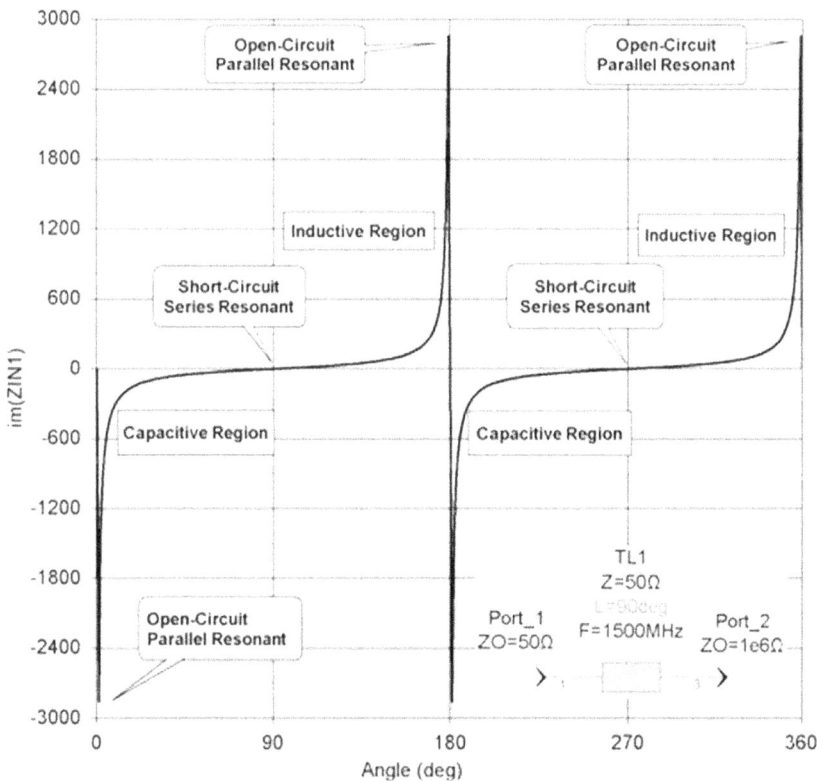

Figure 2-10 Reactance of open circuit transmission line

2.8 Modeling Open-Circuited Microstrip Lines

Care must be used when modeling the open circuit microstrip line due to the radiation effects from the end of the transmission line. The E fields that exist in the air space of the microstrip line add capacitance to the microstrip transmission line. On an open circuit microstrip line this fringing capacitance is referred to as an end effect. The end effect makes the line electrically longer than the physical length. This requires that the physical line length be shortened to achieve the desired reactance.

Calculate the input impedance of a quarter wave open-circuited microstrip transmission line using termination with end effects.

Figure 2-11 shows the correct method of modeling a microstrip
t. As this Figure shows, the impedance of a quarter-wave section of open circuit line is quite low, close to a short circuit. This type of line section could be used as a series resonant circuit.

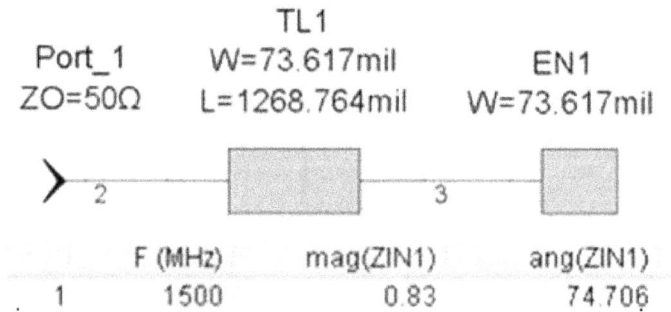

	F (MHz)	mag(ZIN1)	ang(ZIN1)
1	1500	0.83	74.706

Figure 2-11 Quarter wave open circuited line schematic and impedance

2.9 Microstrip Inductive and Capacitive Elements

For short lengths of high impedance transmission line use the following equation to calculate the length of microstrip line to synthesize a specific value of inductance

$$Inductive\ Line\ Length = \frac{f\,\lambda_g\,L}{Z_L}$$

$$Capacitive\ Line\ Length = f\lambda_g Z_C C$$

Example 2-7: Convert the lumped element capacitors and inductors to distributed elements.

Solution: Figure 2-12 shows the low and high impedance microstrip equivalent circuit to the lumped element circuit. The PCB layout shows the line width relationship among the microstrip lines.

The Genesys Advanced TLINE calculator is quite useful to convert an ideal transmission line element to a physical microstrip line. Figure 2-12 shows the lumped element equivalent circuit to the cascade of a low impedance and high impedance microstrip line. The printed circuit layout shows the line width relationship among the 50, 20, and 80 Ohm microstrip lines.

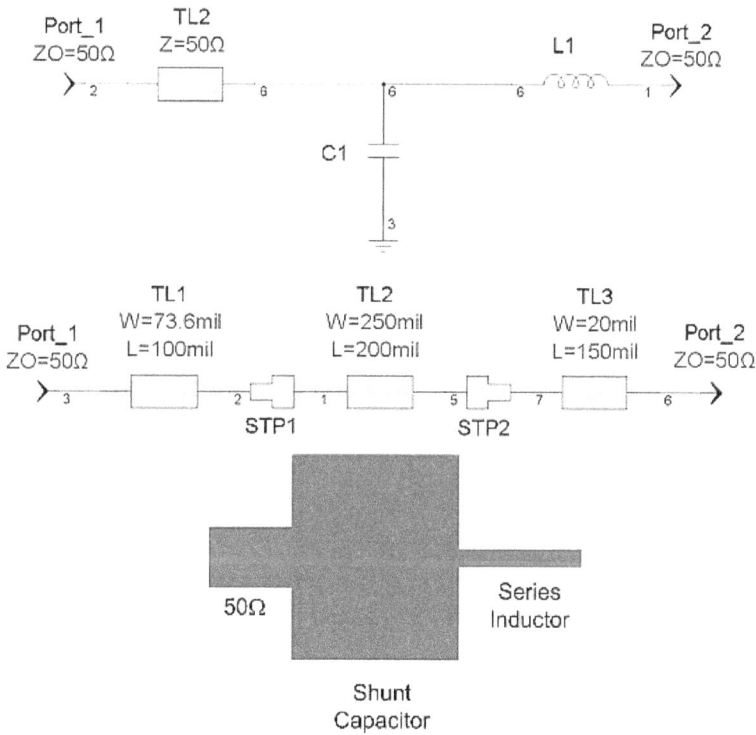

Figure 2-12 Distributed capacitive and inductive lines with PCB layout

2.10 Microstrip Bias Feed Networks

Often it is necessary to insert voltage and current to a device that is attached to a microstrip line. Such a device could be a transistor, MMIC amplifier, or diode. The basic bias feed consists of an inductor and shunt capacitor. At lower RF frequencies these networks are almost entirely realized with lumped element components. Fig. 2-13 shows a typical series inductor, shunt capacitor, lumped element bias feed and its effect on a 50 Ω transmission line.

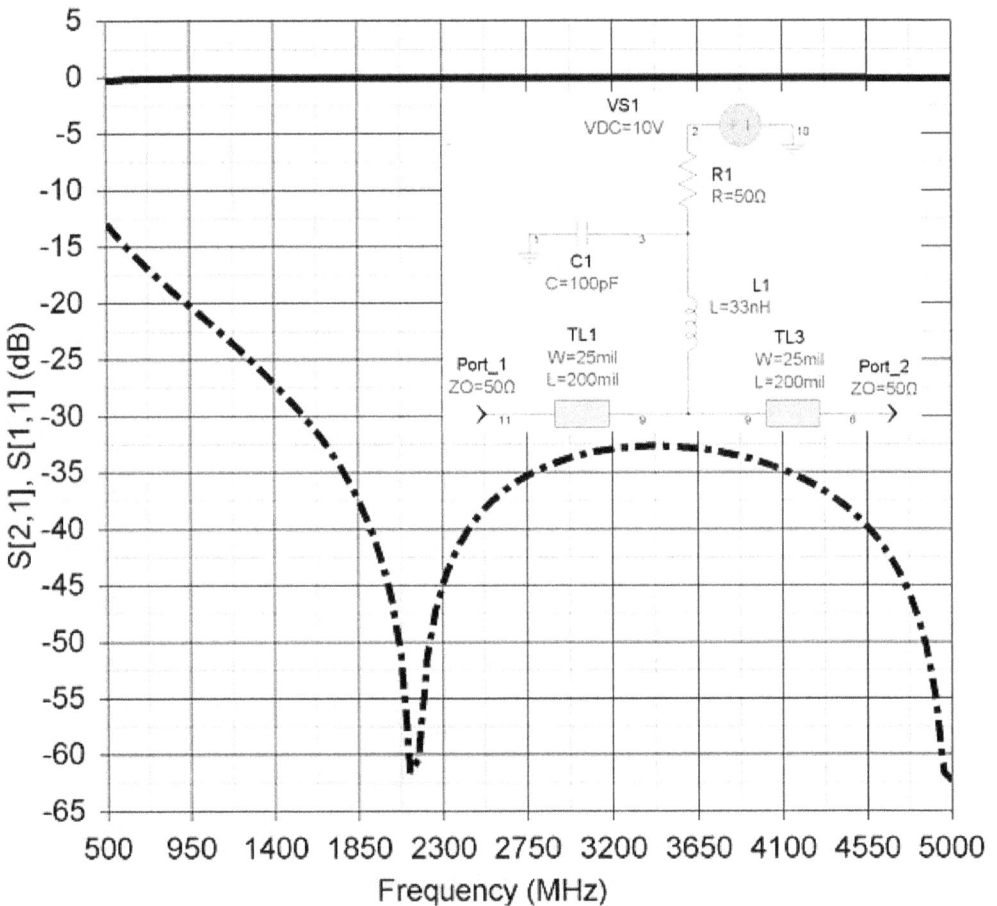

Figure 2-13 Inductor and bypass capacitor bias insertion network

2.11 Distributed Bias Feed

A high impedance microstrip line of $\lambda_g/4$ can be used to replace the lumped element inductor. Similarly a $\lambda_g/4$ of low impedance line can be used to model the shut capacitor.

Example 2-8: Use the Advanced TLine Utility in Genesys to calculate the line length of the $\lambda_g/4$ sections of 80 Ω and 20 Ohm microstrip lines at 2 GHz. Create the schematic of a distributed bias feed network.

Solution: Use the 80 Ω high impedance quarter wave section and a shunt capacitance as shown in Figure 2-14. A microstrip taper, TP1, is used to connect the low impedance line to the high impedance line. Note the use of the microstrip tee junction, TE2. The tee junction accurately models the electrical length of the junction and includes all parasitic effects of the discontinuity. An end-effect element is used on the open circuit line. The return loss null occurs at 1.85 GHz suggesting that the high impedance line length should be decreased to center the design on 2 GHz.

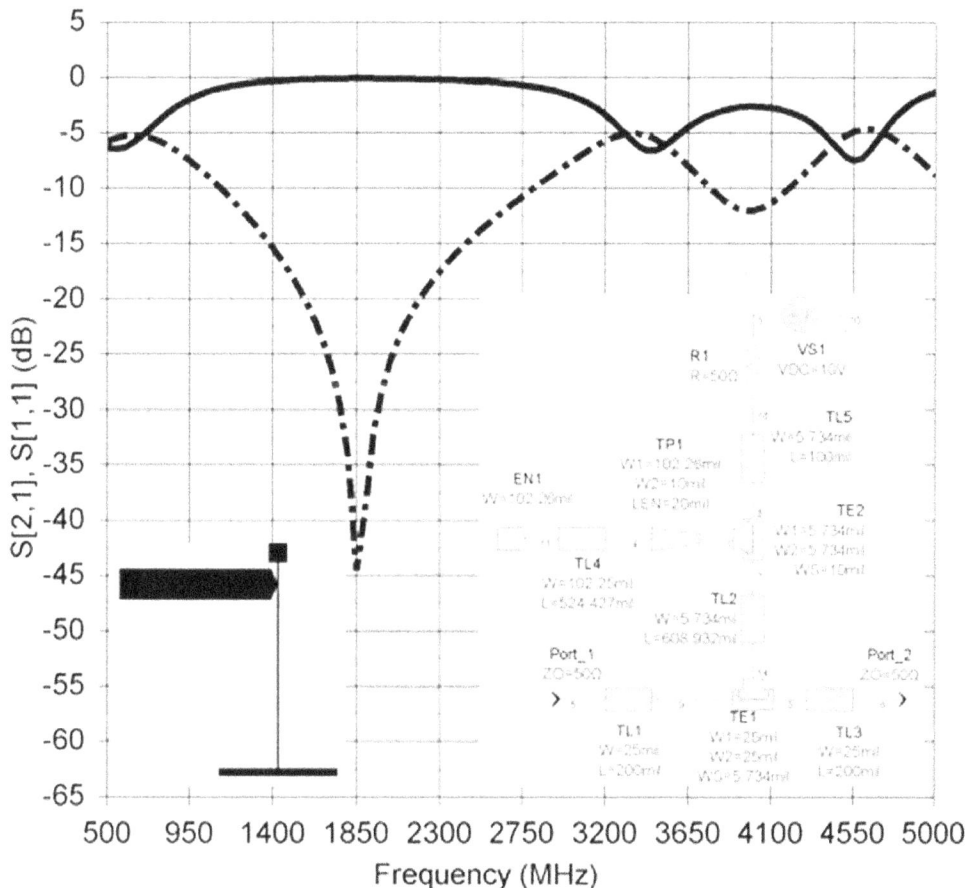

Figure 2-14 Bias feed modeled with distributed transmission line elements

A modified version of the open circuited transmission line is the radial stub. The radial stub is used in applications where an open circuit.

transmission line is needed. Fig. 2-15 shows the use of the radial stub replacing the open circuit transmission line in the bias feed. Comparing the responses we can see that the network using the radial stub achieves a slightly wider bandwidth. The radial stub may also result in a slightly smaller pattern. The PCB patterns are shown in Figs. 2-15.

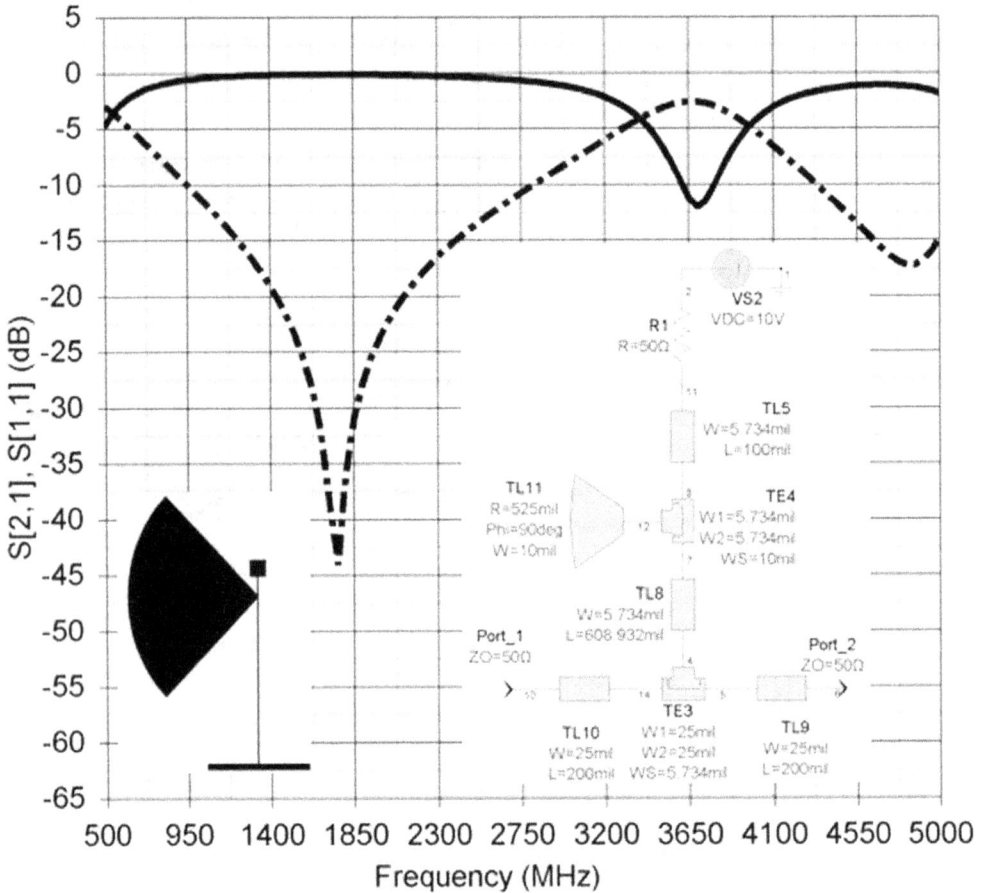

Figure 2-15 Bias feed with open circuited line replaced with the radial stub

2.12.Microstrip Edge Coupled Directional Coupler

Example 2-9: Design an edge coupled microstrip directional coupler with a coupling factor of 10 dB at 5 GHz. Use Rogers RO3003 substrate with relative dielectric constant $\varepsilon_r = 3.0$ and 0.020-inch thickness.

Solution: Traditional coupler design required the computation of the even and odd mode impedance based on the characteristic impedance and coupling factor. Select the coupled microstrip line calculator from the Rectangular transmission lines in the TLINE utility. Enter the dielectric constant, the dielectric thickness, and conductor thickness (t). Select the mode and choose to synthesize a coupled line pair basedon the input characteristic impedance, Z_o, and the coupling in dB. The coupled line width, W, is then calculated to be 38.86 mils and the line spacing, s, is 6.16 mils.

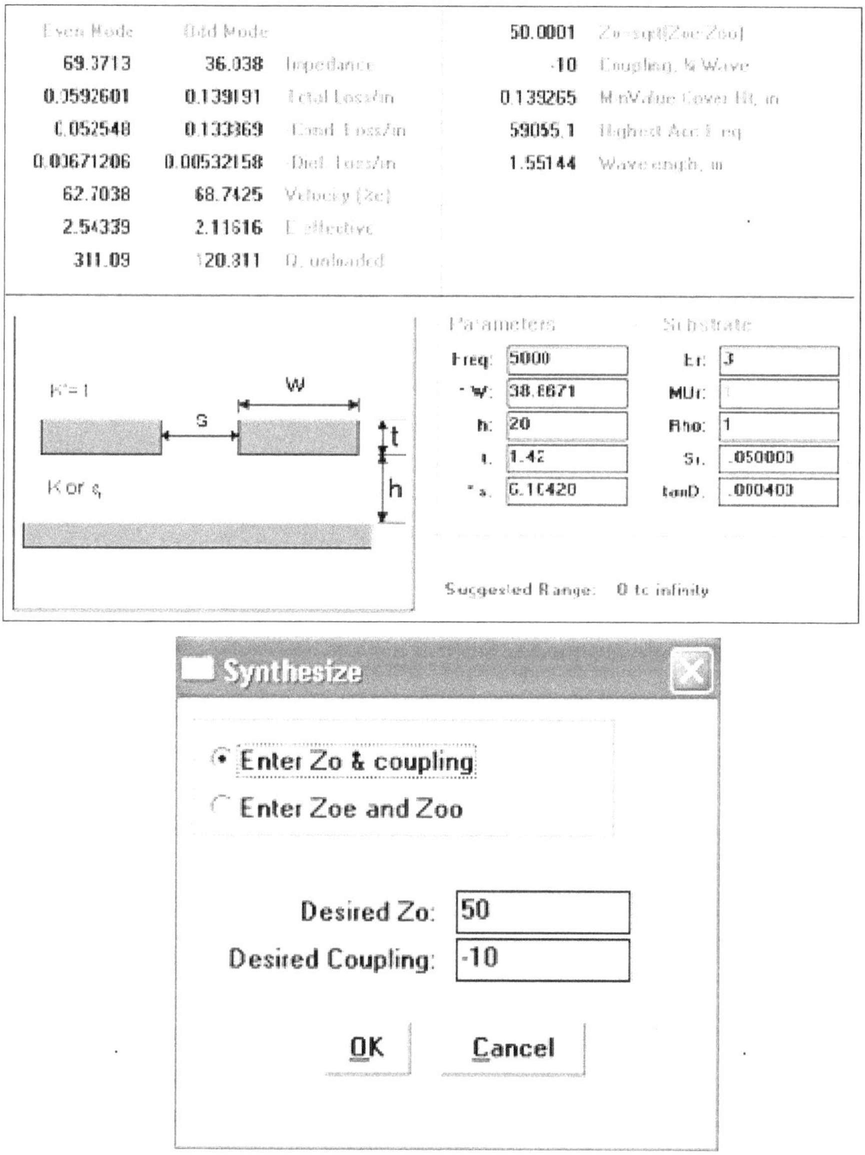

Figure 2-16 TLINE synthesis of a 10dB coupled line section

Add the RO3003, 0.020-inch thick substrate and create the coupler schematic using the Coupled Microstrip Line (Symmetrical) element. On the coupled port side add a Microstrip Bend with Optimal Miter at 90° from the main path. It is important to keep this side orthogonal from the main path so that no further parallel line coupling can occur. The optimal miter element automatically optimizes the miter for the least amount of discontinuity traversing the 90° bend. Lastly add a short section of line to each port to complete the circuit as shown in Figure 2-17. Use a Linear Analysis to sweep the coupler from 4500 MHz to 5500 MHz. Simulate the circuit and display the insertion loss (S21), coupled response (S31), isolation (S41), and the directivity. It is convenient to implement simple mathematical operations directly in the Graph Properties window as opposed to using an Equation block. The directivity is calculated by subtracting the coupled port response from the isolation.

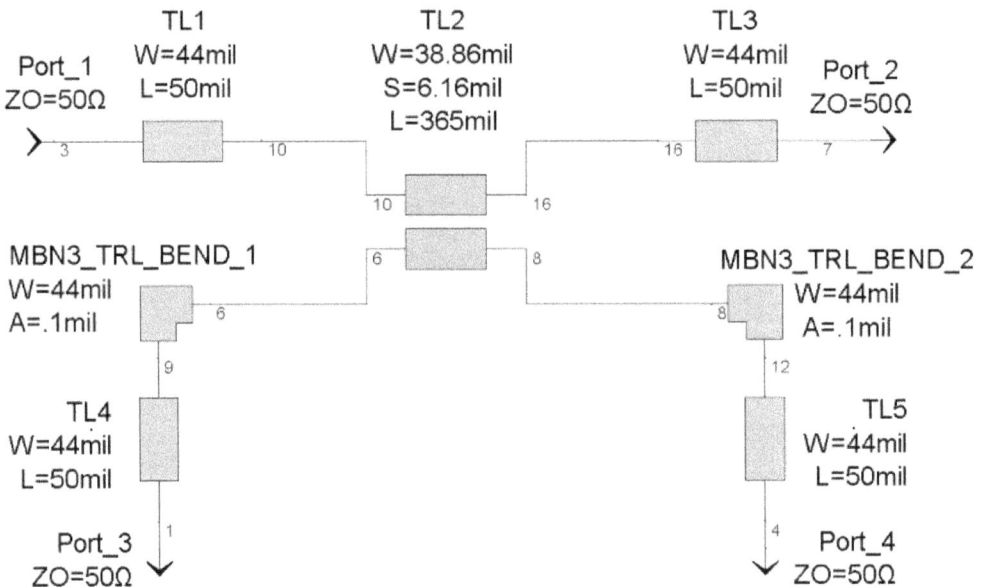

Figure 2-17 Schematic of the microstrip directional coupler

The simulated coupling factor is 10.157 dB which is close to the 10 dB design goal. The isolation is about 25 dB and directivity is about 15 dB. Clearly this coupler is not appropriate for VSWR measurement. It is useful for obtaining a sample of the input signal without disturbing or loading down the input signal.

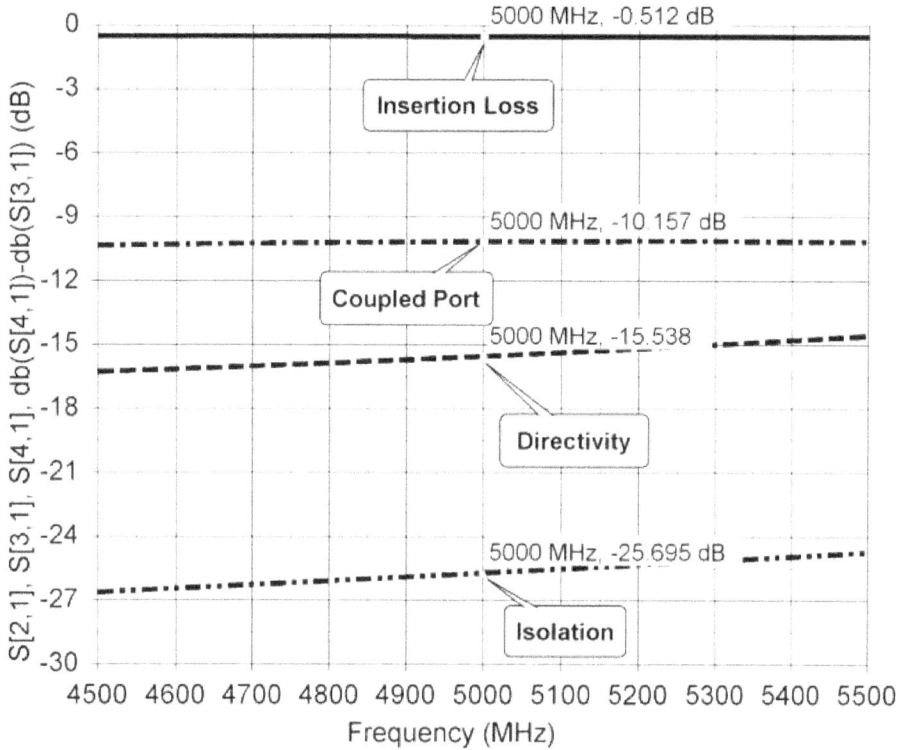

Figure 2-18 Simulated microstrip directional coupler response

References and Further Readings

[1]   David M. Pozar, *Microwave Engineering*, Fourth Edition, John Wiley and Sons, Inc., 2012

[2]   Ali Behagi and Stephen D. Turner, *Microwave and RF Engineering,* A Simulation Approach with Keysight Genesys Software, BT Microwave LLC, March, 2015

[3]   Keysight Technologies, Genesys , Users Guide, www.keysight.com.

[4]   William Sinnema and Robert McPherson, Electronic Communications, Prentice-Hall Canada, Inc., Scarborough, Ontario, 1991

[5]   UHF/Microwave Experimenters Manual, American Radio Relay League, Newington, CT.1990

[6]   Reference: I. J. Bahl and D. K. Trivedi, "A Designer's Guide to Microstrip Line", Microwaves, May 1977, pp. 174-182.

[7]   Microwave Handbook Volume 1, Radio Society of Great Britain, The Bath Press, Bath, U.K., 1989.

[8]   Tatsuo Itoh, Planar Transmission Line Structures, IEEE Press, New York, NY, 1987

Problems

2-1.   The input reflection coefficient of a transistor is measured to be 0.22 at an angle of 32°. Determine the input VSWR of the device.

2-2.   Determine the VSWR of a sattelite antenna with a return loss of -11.4 dB.

2-3.   Determine the impedance of a quarter-wave transformer to match a 25 Ω load to a 50 Ω source. Design the quarter-wave transformer from Problem 3 using a microstrip transmission line. The frequency of operation is 2.05 GHz. The dielectric constant is 3.0 with a thickness of 0.030 in. Determine the length and width of the microstrip line.

2-4.   A radio transmitter is operating into a transmission line that measures a 3:1 VSWR. Determine the percentage of power that would be expected to reflect back into the transmitter.

2-5.   A series RLC load, R = 75 Ω, L = 10 nH, C = 25 pF is connected to a 50 Ω transmission line. Setup a Linear Analysis in Genesys to sweep the frequency from 200 MHz to 2000 MHz in 200 MHz steps. Display the input reflection Coefficient, S11, and VSWR in a Table.

2-6.   Create a simple schematic using the RG8 coaxial cable. Set the length to 50 ft. Calculate the insertion loss in a Table. Terminate the coaxial line with a 100 Ω resistor and display the input return loss and reflection coefficient in the same Table.

2-7.   Calculate the cutoff frequency of the TE1,0 mode in a rectangular waveguide with a height of 0.200 inches and a width of 0.470 inches. Also calculate the waveguide wavelength, λ_g.

2-8.   Design a distributed bias feed network for a C Band amplifier operating at 6.0 GHz. Use a microstrip substrate with a dielectric constant of 10.2 and a thickness of 0.025 inches. Plot the insertion loss and return loss from 2 GHz to 10 GHz.

2-9.   Determine the physical length of a $\lambda_g/4$ open circuit microstrip transmission line with an impedance of 20 Ω. The frequency of operation is 10 GHz. Use a microstrip dielectric constant of 2.2 and a thickness of 0.010 inches. Determine whether an end-effect model element should be used.

2-10.   Design a 35dB directional coupler to be used in a 100W transmitter operating at 4 GHz. Design the coupler in stripline. Use a dielectric constant of 3.0 and a thickness of 0.060 for each half of the stripline transmission media. Using Genesys determine the directivity of the coupler. Comment on coupler's ability to measure a 1.25:1 VSWR.

Chapter 3

Network Parameters and the Smith Chart

3.1 Introduction

At low frequencies below the VHF range, the terminal voltages V_1 and V_2 and the terminal currents I_1 and I_2 of a two-terminal network, shown in Fig. 3-1, can be related to each other by a different set of matrix parameters. The most common representations are the impedance matrix (Z parameters), the admittance matrix (Y parameters), the hybrid matrix (h parameters), and the transmission matrix (ABCD parameters).

Figure 3-1 Low frequency two-port network

Plot the impedance $Z = 25 + j25\ \Omega$ on the standard Smith Chart.

Solution: To display the impedance on the Smith chart, create a new workspace in Genesys with the impedance and simulate the schematic. Select the Smith chart in data display window.

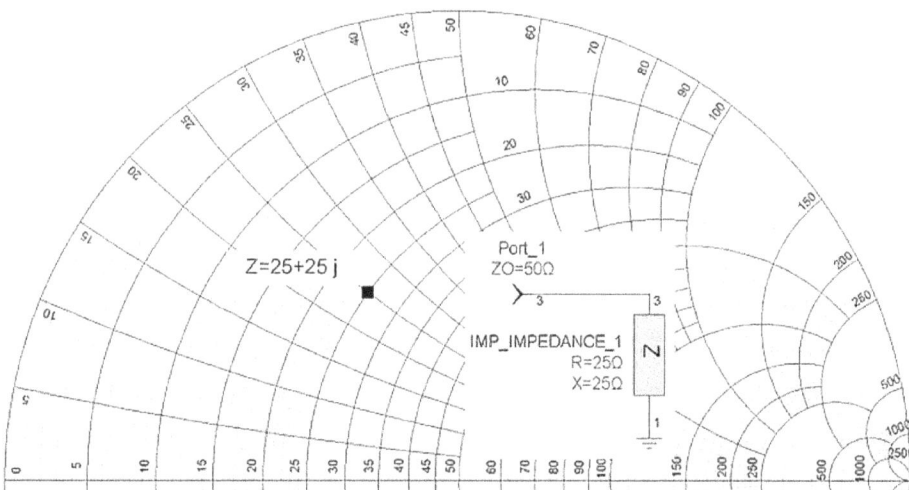

Figure 3-2 Plotting impedance on the Smith Chart

3.2 Lumped Element Movements on the Smith Chart

Lumped element movements on the Smith Chart form the basis for impedance matching. The Smith Chart is a wonderfully intuitive tool, for the visualization of moving from one impedance to another, without involving circuit synthesis mathematics. Understanding the basic movements around the Smith Chart will build a foundation for the circuit designs covered in this text. It is helpful to display both the impedance and admittance coordinates simultaneously on the Smith Chart.

Example 3-1: Measure the amount of reactance required to move the impedance Z = 25 + j25 Ω from point A to point B on the Smith Chart, as shown in Figure 3-3.

Solution: The amount of reactance required in the inductor can be measured from the reactance lines that intersect the start point (A) and end point (B). As Figure 3-3 shows the reactance is approximately 0.364 Ω. Using a design frequency of 1000 MHz the inductance is calculated to as 2.86 nH.

$$L(series) = \frac{0.364 \cdot (50)}{2 \cdot \pi \cdot 1000 \cdot 10^6} = 2.86 \ nH$$

Add the inductor to the impedance element in Genesys to verify that a 2.86 nH inductor moves the impedance to point B on the Smith Chart. Make the inductor tunable and set the initial value to 0.1 nH and tune the value of inductance to move the impedance from point A to point B.

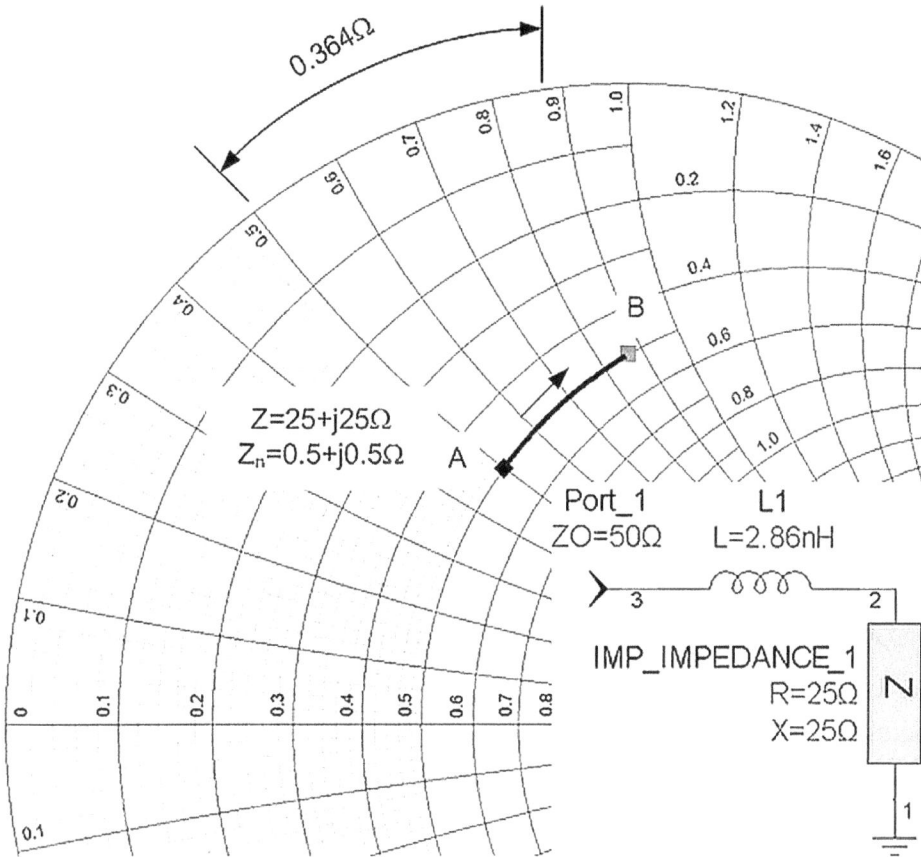

Figure 3-3 Series inductance added to 25 + j25 Ω impedance

3.3 Adding a Shunt Reactance to an Impedance

Adding a shunt element to an impedance point on the Smith Chart causes the resulting impedance to move along the constant conductance circle in which the impedance intersects. A shunt inductance will move the impedance in a counter-clockwise direction while a shunt capacitance will move the resulting impedance in a clockwise direction on the constant conductance circle. The susceptance that is added to the impedance by the shunt element can be read from the Smith Chart by finding the difference between the lines of susceptance that intersect the start point and end point on the constant conductance circle.

Example 3-2: Measure the susceptance required to move from point B to point C on the real axis of the Smith Chart.

Solution:. Add a shunt capacitance to move the impedance from point B to point C as shown in Figure 3-4. When adding a shunt element switch from the impedance grid to the admittance coordinates. The admittance follows the constant conductance circle in which the point lies by the difference between the intersecting susceptance lines. The susceptance as measured on the perimeter of the chart is 0.865 mhos. Calculate the capacitance.

$$C\ (shunt) = \frac{B}{\omega n} = \frac{0.865}{2 \cdot \pi \cdot 1000 \cdot 10^6 \cdot (50)} = 2.75\ pF$$

Alternatively you can add a shunt capacitor to the circuit and make the capacitance value tunable. Start with a very low value of approximately 0.1 pf and increase the value of capacitance until the admittance is moved from point B to point C.

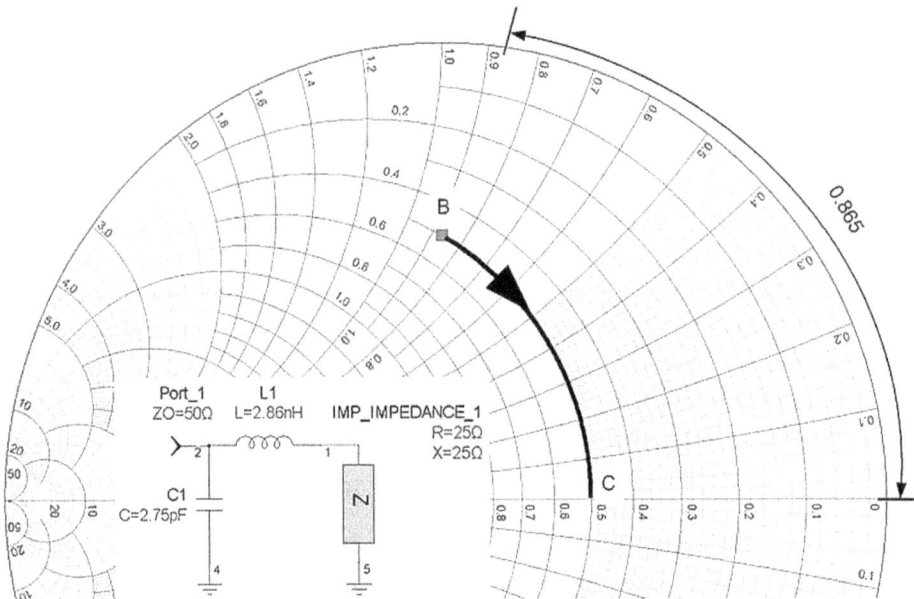

Figure 3-4 Shunt capacitance added to the network

Example 3-3: Show the direction of movement on the Smith chart when adding a series or shunt element to an impedance.

Solution: By iterative addition of series and shunt elements an impedance point can be moved from one location to another on the Smith chart at a given frequency.

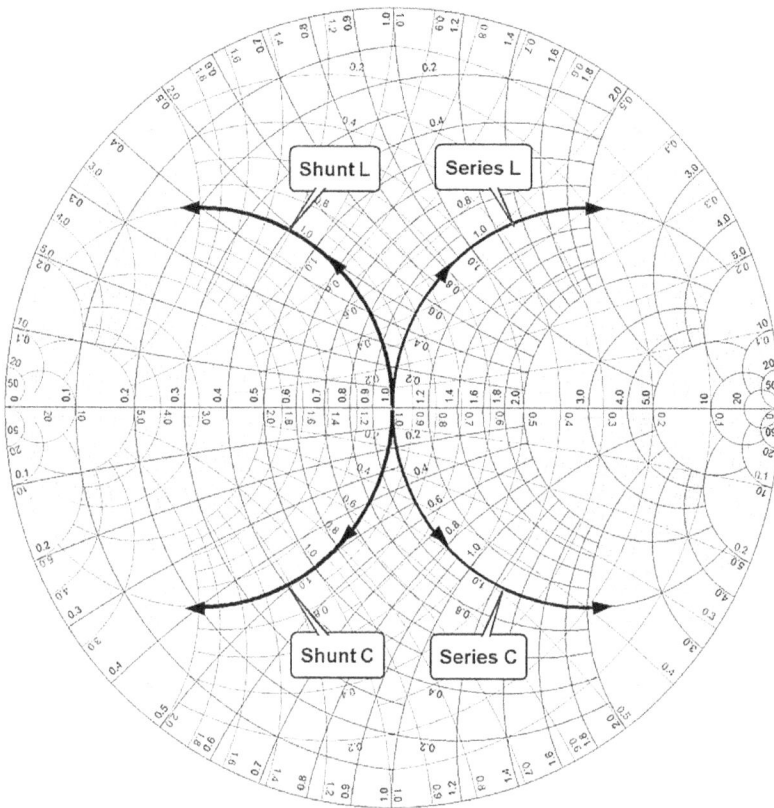

Figure 3-5 Lumped element movements on the Smith Chart

3.4 VSWR Circles on the Smith Chart

It has been shown that the VSWR of a network is related to the magnitude of the reflection coefficient, independent of the angle. From plotting the reflection coefficient on the Smith Chart we know that the origin of the vector is located in the center of the chart. This suggests that as the reflection coefficient vector rotates 360 degrees around the chart with a constant magnitude, the VSWR will remain constant. This locus of points around the center of the Smith Chart is known as the constant VSWR circle

Plot the VSWR circles on the Smith Chart for VSWR of 1.25, 2, 4, and 10.

Create a new workspace and open a new schematic to analyze the network at a fixed frequency as shown in Figure 3-6.
Any frequency can be entered as the model uses a fix frequency.

The constant VSWR circles can be displayed in Genesys by using the manual S parameter model element and a Parameter sweep. The manual S parameter model is shown in Figure 3-6. In this Figure, the impedance or reflection coefficient will not change with frequency. The magnitude of S_{11}, which is the magnitude of the reflection coefficient, is assigned a variable named mag1. An Equation Editor block can be used to enter the desired VSWR and calculate the reflection coefficient. The reflection coefficient is then assigned to the variable "mag11".

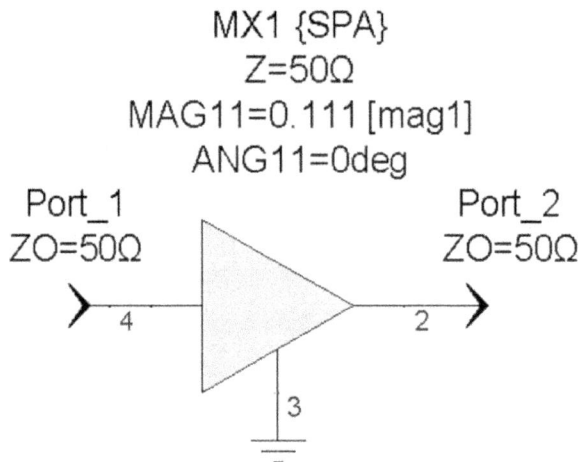

MX1 {SPA}
Z=50Ω
MAG11=0.111 [mag1]
ANG11=0deg

Port_1   Port_2
ZO=50Ω   ZO=50Ω

4   2

3

Figure 3-6 Manual S Parameter model in Genesys

Make the angle of S_{11} on the S parameter model a tunable variable. Then create a parameter sweep that will be used to sweep the angle of S_{11} from 0 to 360 degrees in 2 degree steps. Create a Linear Analysis to analyze the network at a fixed frequency as shown in Figure 3-7. Any frequency can be entered as the model is frequency independent. The equation editor block is presented in Figure 3-8. Enter the desired VSWR value in the Equation block and simulate the circuit. A family of circles is shown in Figure 3-7 for VSWR values of: 1.25, 2.0, 4.0, and 10.0. Any reflection coefficient intersecting the circle has a VSWR equal to the value of the circle. If the reflection coefficient is inside the circle, the VSWR is less than the value of the circle. Likewise if a reflection coefficient is outside of the circle, the VSWR is greater than the value of the circle. Figure 3-7 gives intuitive feel for relating reflection coefficients on the Smith Chart to their VSWR value.

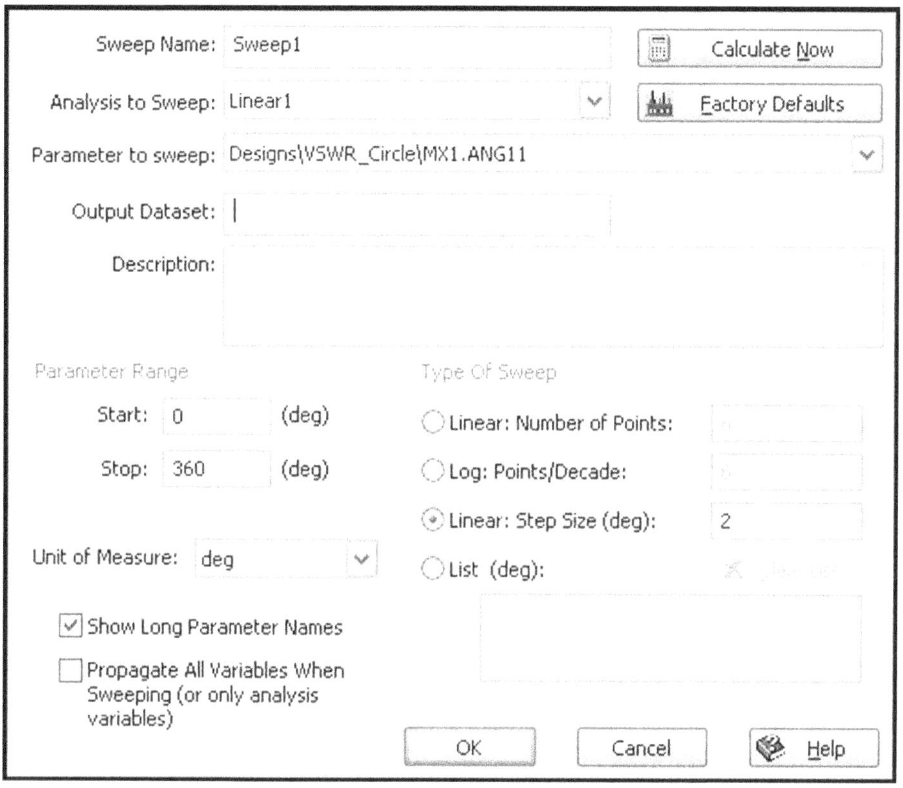

Sweep Name: Sweep1	▦ Calculate Now
Analysis to Sweep: Linear1 ∨	⚒ Factory Defaults
Parameter to sweep: Designs\VSWR_Circle\MX1.ANG11	∨
Output Dataset: |	
Description:	

Parameter Range

		Type Of Sweep
Start: 0 (deg)		○ Linear: Number of Points:
Stop: 360 (deg)		○ Log: Points/Decade:
		⦿ Linear: Step Size (deg): 2
Unit of Measure: deg ∨		○ List (deg):

☑ Show Long Parameter Names

☐ Propagate All Variables When Sweeping (or only analysis variables)

OK	Cancel	🛠 Help

Figure 3-7 Parameter sweep for the VSWR circle

```
1    'Enter Sweep Parameters
2    vswr1=1.25
3    mag1=(vswr1-1)/(vswr1+1)
4    angle=Sweep1_Data.MX1_ANG11SWP_F
5    Gamma1=Sweep1_Data.S(1,1)
6    =setindep("Gamma1","angle")
7
```

Figure 3-8 Generation of the VSWR circle on the Smith Chart

From the plot of the VSWR circles in Figure 3-9 we can see that the value of the VSWR is equal to the magnitude of the reflection coefficient as labeled on the horizontal axis to the right of the center of the chart. Conversely the horizontal axis values on the left-hand side of the chart represent $1/\Gamma$.

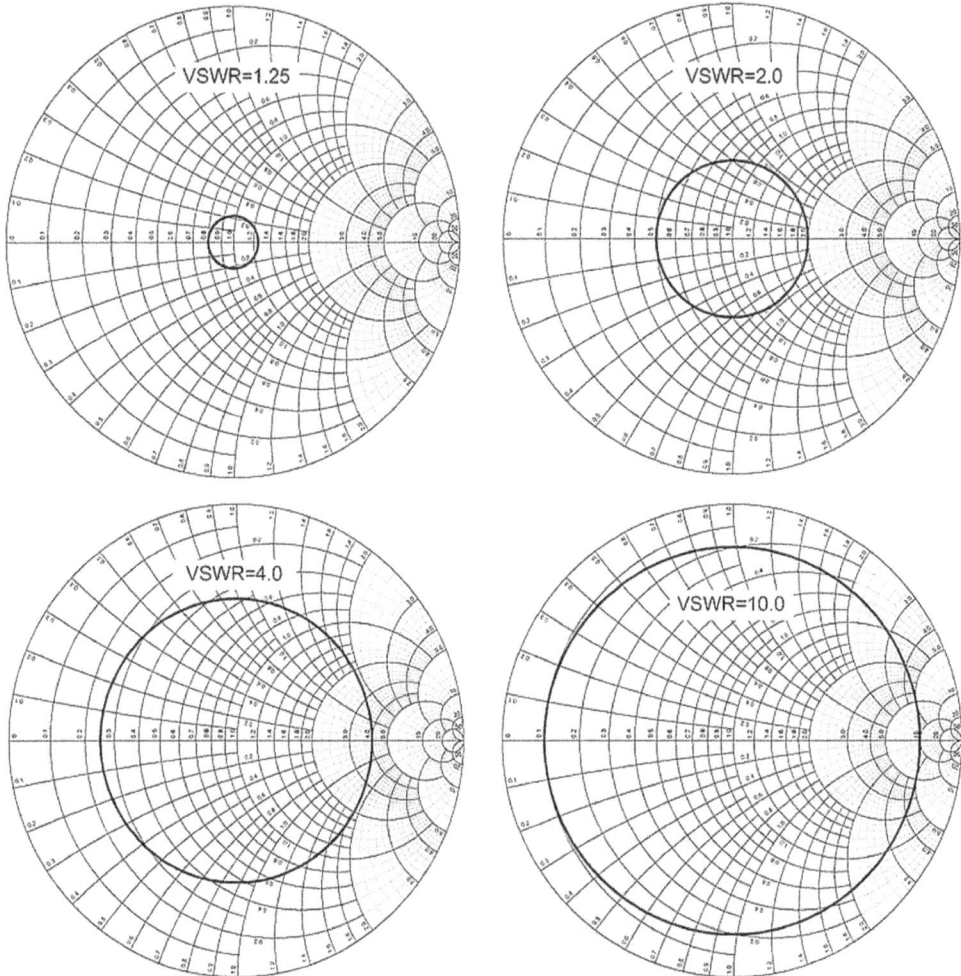

Figure 3-9 VSWR circles for VSWR values of 1.25, 2.0, 4.0, and 10.0

3.5 Adding a Series Transmission Line to Impedance

We have seen that adding a reactance in series with an impedance point causes the impedance to follow the constant resistance circles. Adding a transmission line of the same impedance as the Smith Chart's normalized impedance, in series with an impedance point causes the resulting impedance to follow the constant VSWR circle in which the impedance lies. The impedance moves in a clockwise direction on the constant VSWR circle.

Example 3-4: Revisiting the 25+j25 Ω impedance in Genesys, calculate calculate the electrical length of a series transmission line moving the impedance at point A to point B on the Smith Chart, as shown in Figure 3-10.

Solution: Add a series transmission line to the impedance and make the electrical length tunable. Increase the line length to move the impedance to point B. The electrical line length is 58.45 degrees. The impedance on the real axis (zero reactance) on the right hand side of the Smith Chart would represent a point of maximum voltage-minimum current along the transmission line.

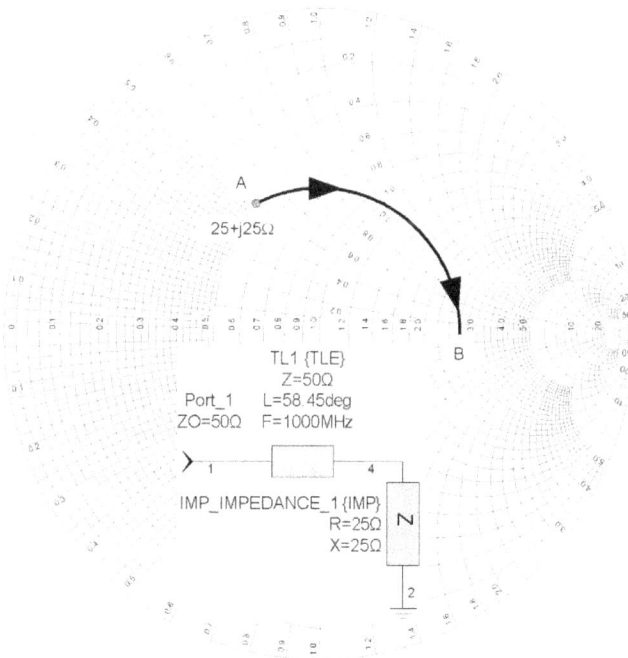

Figure 3-10 Series transmission line moves impedance to point B

Transmission line lengths are sometimes referred to in terms of fractional wavelengths. Because one wavelength is equal to 360°, a 58.45° electrical length represents (58.45/360) 0.162λ.

Example 3-5: Calculate the electrical length of a series transmission line moving the impedance at point B to point C on the Smith Chart.

Solution: Continue to add electrical length to the transmission line to reach point C on the Smith Chart. The real impedance (zero reactance) on the left side of the horizontal axis on the Smith Chart represents a minimum voltage-maximum current point along the transmission line.

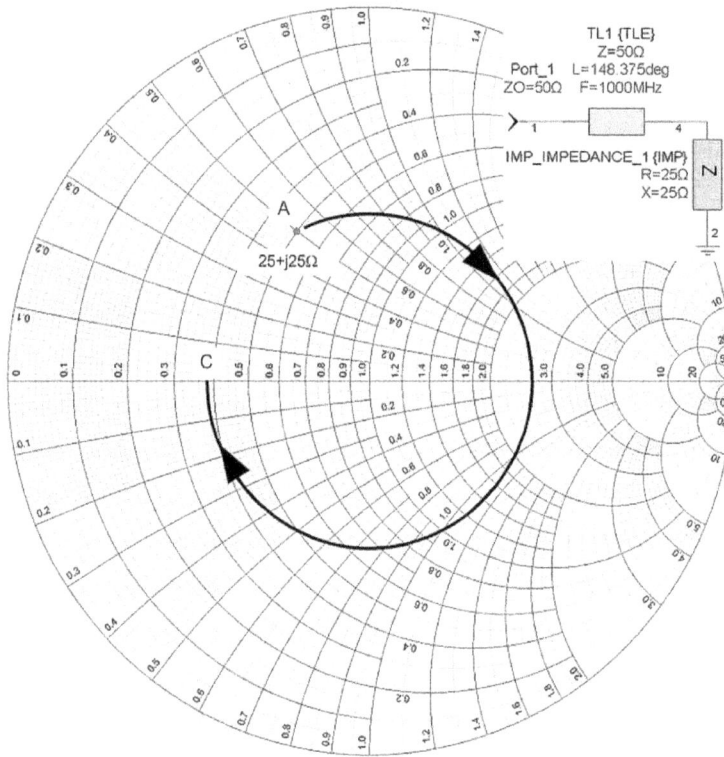

Figure 3-11 Series transmission line moves impedance

Further increasing the length of the transmission line we find that we arrive back at point A at 180 degrees of electrical length. Therefore the electrical distance around the Smith Chart is 180° or λ/2 wavelength. The points of maximum voltage and minimum voltage will repeat every λ/2 wavelength.

3.6 Adding a Shunt Transmission Line to Impedance

We have seen that the open and short-circuited transmission lines could take on the equivalence of an inductor, capacitor, or series and parallel resonant circuits depending on the electrical length of the line. Therefore the shunt transmission line will behave more like the lumped element movements on the Smith Chart.

3.7 Open and Short Circuit Shunt Transmission Lines

At DC and low frequencies, a short circuit is a very low inductance but this is not the case at higher RF and microwave frequencies. Figure 3-12 shows that a transmission line with 0° length (perfect short) appears at the short circuit point on the Smith Chart. As 45° of electrical length is added to the short circuit, the impedance moves clockwise along the outer circumference of the Smith Chart to the position shown at the top of the chart. As the line length is increased to 90° we see that the short circuit has been transformed to an open circuit. At 180° line length the impedance will travel completely around the Smith Chart and appear as a short circuit again. Therefore depending on the line length the short circuit transmission line can be transformed into a shunt capacitor or shunt inductor.

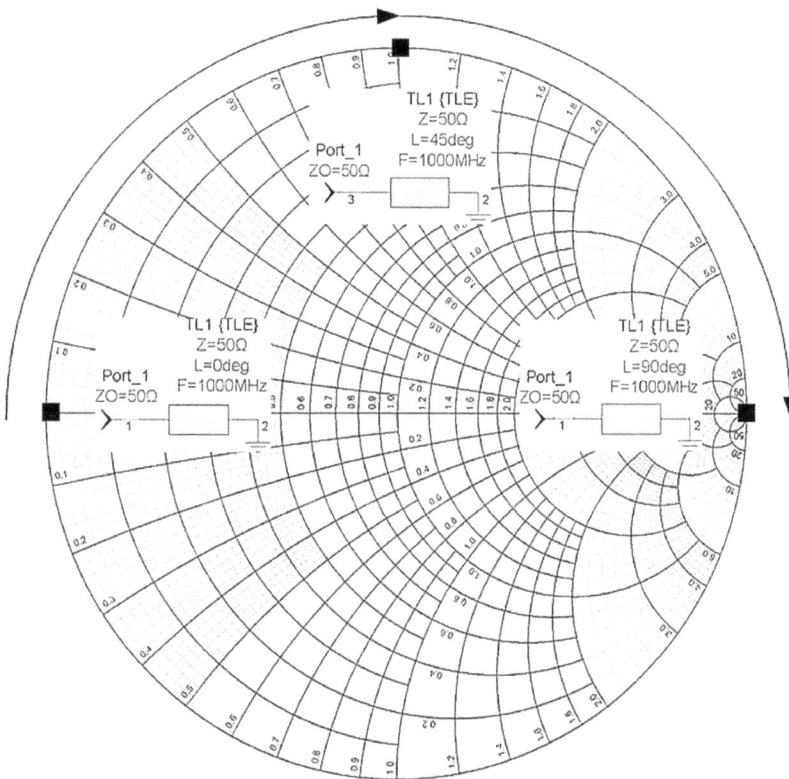

Figure 3-12 Short circuit impedance at electrical lengths: 0°, 45°, and 90°

Figure 3-13 shows that the characteristic of the open circuit transmission line. At 0° electrical line length it appears as a perfect open circuit. As the electrical length is increased, the impedance moves clockwise around the circumference to the 45° position at the bottom of the chart. At 90° electrical length the open circuit now appears as a short circuit. This property of transforming open circuits to short circuits and vice versa is one that is used frequently throughout microwave circuit design.

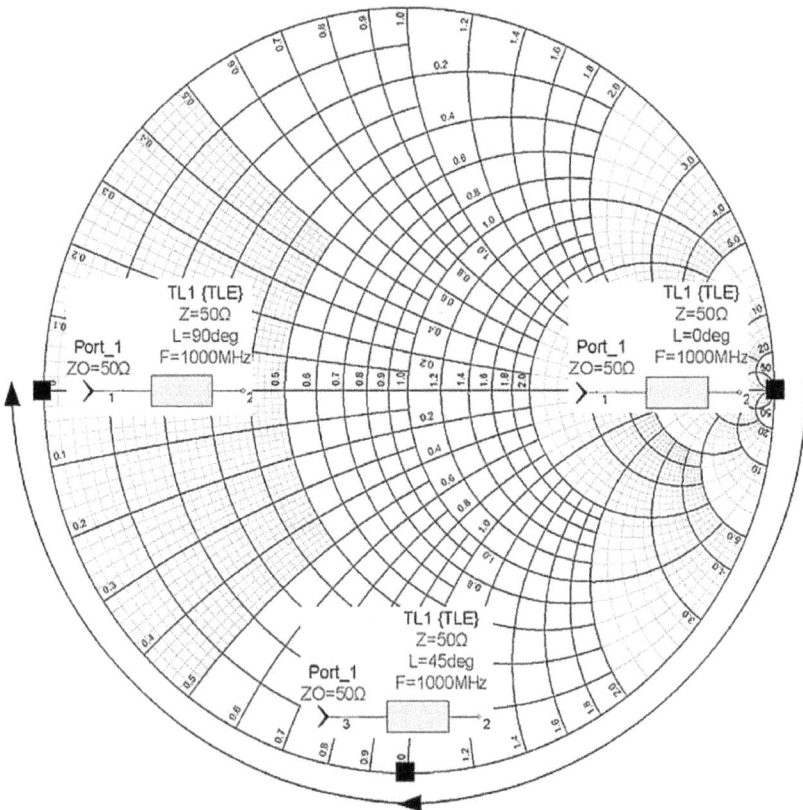

Figure 3-13 Open circuit transmission line impedance at electrical lengths: 0°, 45°, and 90°

For small fractional wavelength transmission lines the open circuit shunt transmission line acts as a shunt capacitor.

Example 3-6: Measure the electrical length of a shunt transmission line to move the impedance from point A to the center of Smith Chart, as shown in Figure 3-14.

Solution: Plotting the impedance on the Smith Chart with the admittance circles shows that the impedance lies directly on the unit conductance circle. Therefore an open circuit shunt transmission line can move this impedance directly to 50 A .Ωs the schematic inset of Figure 3-14 shows, a 45.38° electrical length of a 50 Ω shunt transmission line moves the impedance to the center of the Smith Chart. The second schematic inset of Figure 3-14 shows that this circuit is equivalent to a 3.2 pF capacitor in shunt with the load impedance.

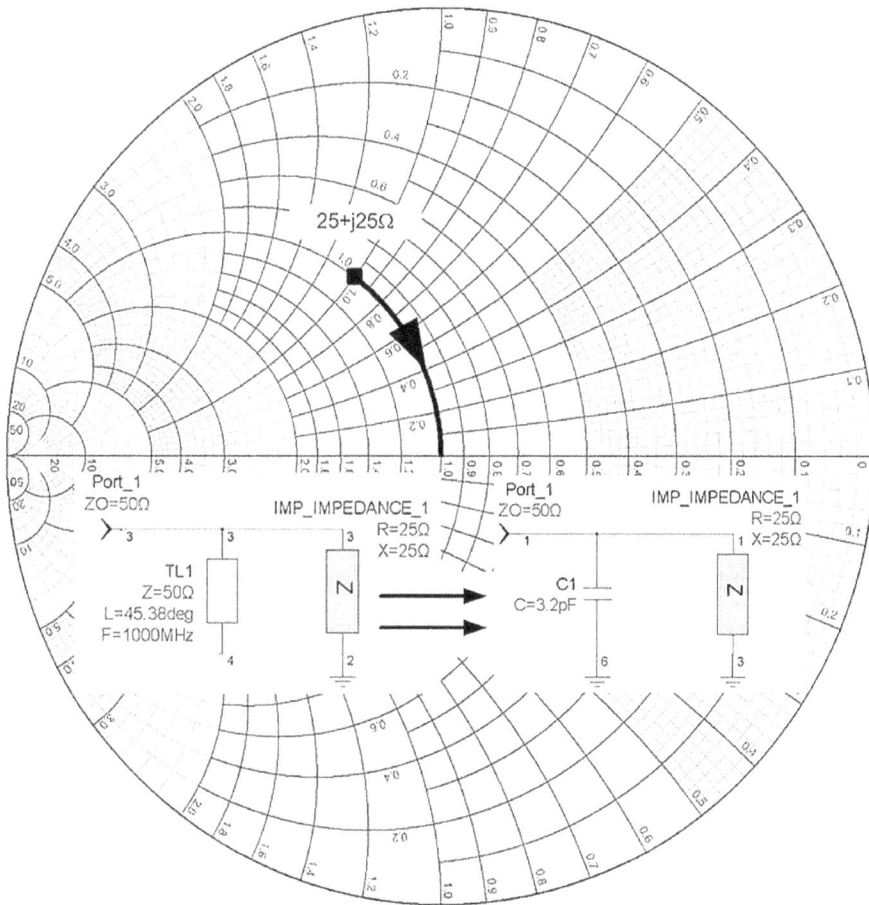

Figure 3-14 Shunt transmission line added to 25+j25 Ω impedance

For small fractional wavelength transmission lines the short circuit shunt transmission line acts as a shunt inductor. Consider the 4.3 –j14 Ω impedance as shown in Figure 3-15. This impedance lies on the unit conductance circle on the bottom half of the Smith Chart. A 50 Ω shunt.

transmission line added to the impedance moves along the constant conductance circle to the center of the Smith Chart. Similarly a 2.4 nH shunt inductor has the same effect at a frequency of 1000 MHz. These movements form the basis for distributed network impedance matching which is covered in detail in Chapter 6.

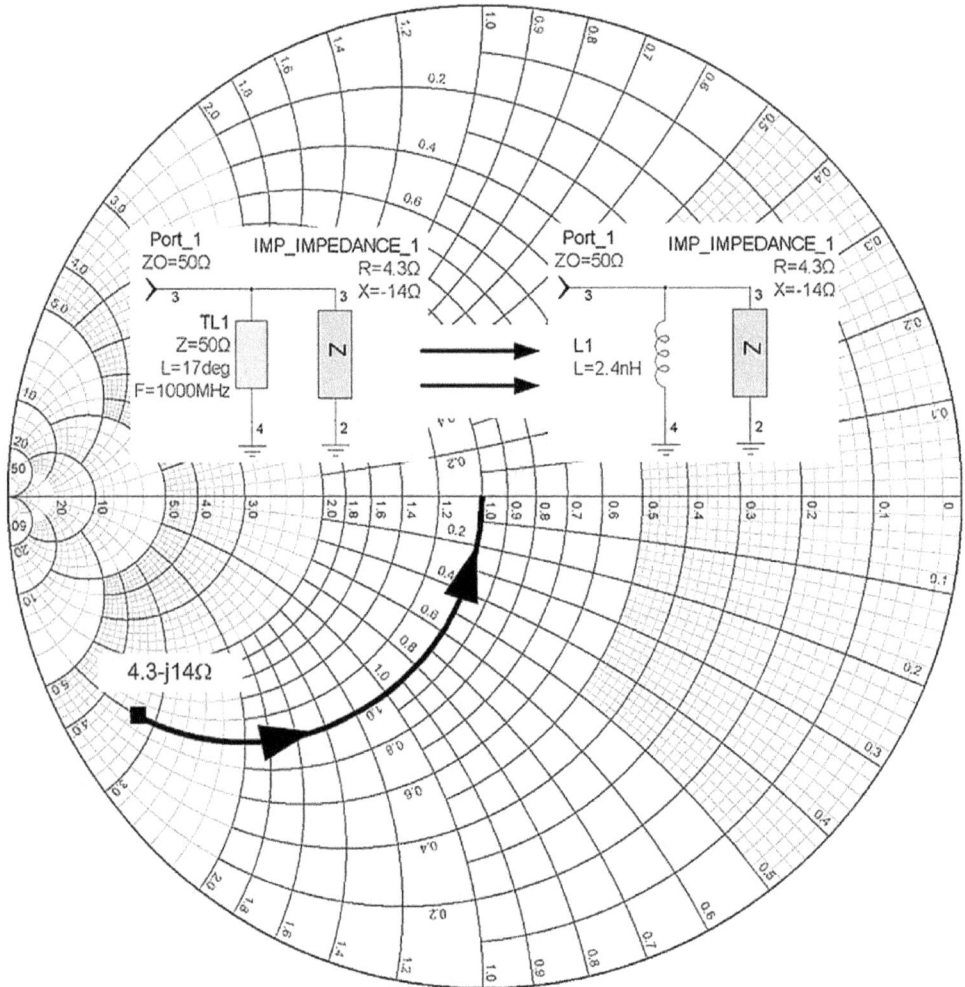

Figure 3-15 Short circuit shunt transmission line added to 4.3-j14 Ohm impedance

References and Further Reading

[1]   David M. Pozar, *Microwave Engineering*, Fourth Edition, John Wiley and Sons, Inc., 2012

[2]   Ali A. Behagi and Stephen D. Turner, *Microwave and RF Engineering*, A Simulation Approach with Keysight Genesys Software, BT Microwave LLC, March, 2015

[3]   William Sinnema, Electronic Transmission Technology, Prentice-Hall, Inc, Englewood Cliffs, New Jersey 07632, 1979

[4]   Chris Bowick, RF Circuit Design, Second Edition, Newnes, Elsevier, 2008

[5]   Keysight Technologies, *Genesys 2015.08, Users Guide*, www.keysight.com

Problems

3-1.   Place a 20 + j30 Ω impedance at point A on the Smith Chart. Add a series inductance to move the impedance along the constant resistance circle to point B having the impedance 20 + j50 Ω. Using a design frequency of 1000MHz, calculate the inductance that this reactance represents.

3-2.   Continuing with Problem 3-1 enable the admittance coordinates on the Smith Chart to add a shunt element. Add a shunt capacitance to move the impedance to the real axis as shown in Figure 3-16. Measure the susceptance required to move to the real impedance axis by the difference between the intersecting susceptance lines.

3-3.   Use the Genesys Equation Editor to calculate the magnitude of reflection coefficient for a desired VSWR = 2.

3-4.   Using Genesys, create a constant VSWR circle for a VSWR=20. Comment on the VSWR value required to place the VSWR circle on the circumference of the Smith Chart.

3-5.   A load of $75 + j20$ Ω is connected to a 50 Ω transmission line. Calculate the load admittance and the input impedance if the line is 0.2 wavelengths long.

3-6.   For the load impedance in Problem 3-4, determine the reflection coefficient and the transmission coefficient.

3-7.   For the load impedance in Problem 3-4, determine the normalized value of the load impedance if the impedance is normalized to 75 Ω.

3-8.   A series RLC load, $R = 100$ Ω, $L = 20$ nH, $C = 25$ pF is connected to a 50 Ω transmission line. Calculate the VSWR and reflection coefficient at the load at 100 MHz.

3-9.   Using the RLC load impedance of problem 3-8 determine the impedance with a series transmission line of characteristic impedance of 50 Ω and electrical length of 180 degrees.

3-10.   Determine the input impedance of a network that has a reflection coefficient of 0.5 at an angle of 112°.

3-11.   Create a one-port S parameter text file with the impedance of Problem 3-10 at a frequency of 1 GHz. Plot the S parameter, S11, on the Smith Chart in Genesys.

3-12.   Using the S parameter file of the C Band amplifier of Figure 3-8, plot the input return loss, S11, and output return loss, S22, on the Smith Chart. Determine the worst case input and output VSWR for this amplifier.

Chapter 4

Resonant Circuits and Filters

4.1 Series Resonant Circuits

In this section we analyze the behavior of the resonant circuit in Genesys.

Example 4-1: Analyze the one port resonator that is represented as a series RLC with R = 10 Ω, L = 10 nH, and C = 10 pF.

Solution: The plot of the resonator's input impedance in Figure 4-1 shows that the resonance frequency is about 503.3 MHz and the input impedance at resonance is 10 ,Ω the value of the resistor in the network.

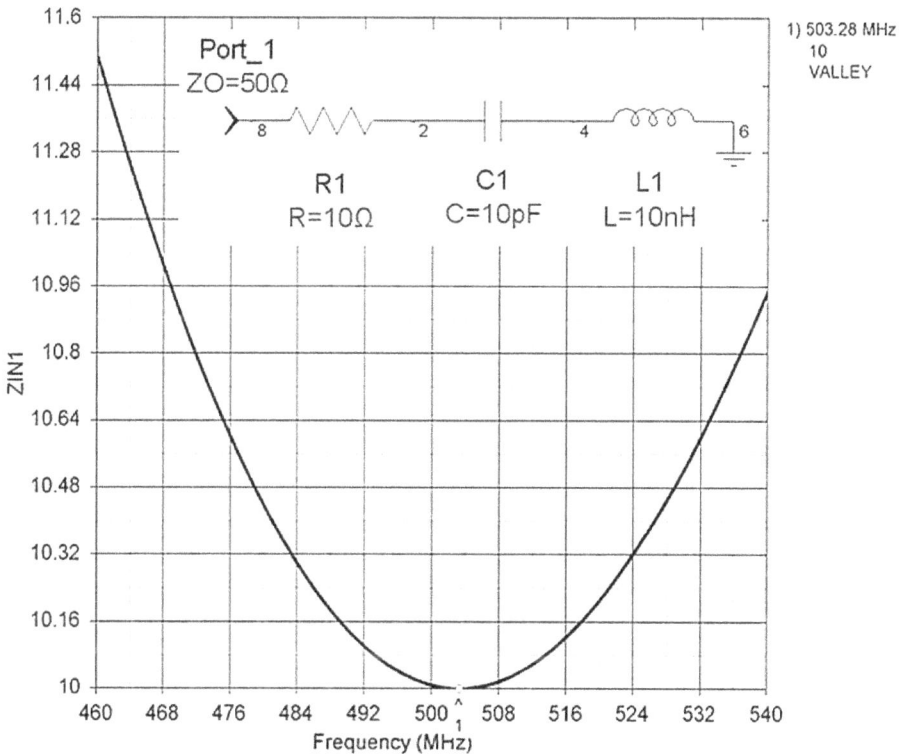

Figure 4-1 One-port series RLC resonator circuit and input impedance

The input impedance of the series RLC resonant circuit is given by,

$$Z_{in} = R + j\omega L - j\frac{1}{\omega C}$$

where, $= 2\pi f$ is the angular frequency in radian per second.

If the AC current flowing in the series resonant circuit is I, then the complex power delivered to the resonator is:

$$P_{in} = \frac{|I|^2}{2} Z_{in} = \frac{|I|^2}{2} \left(R + j\omega L - j\frac{1}{\omega C} \right)$$

At resonance the reactive power of the inductor is equal to the reactive power of the capacitor. Therefore, the power delivered to the resonator is equal to the power dissipated in the resistor.

$$P_{in} = \frac{|I|^2 R}{2}$$

4.2 Parallel Resonant Circuits

Example 4-2: Analyze the parallel RLC resonator of Figure 4-1 with R = 10 Ohm, L= 10 nH and C = 10 pF. The schematic of Figure 4-2 represents the lumped element representation of the parallel resonant circuit.

Solution: The plot of the magnitude of the input impedance shows that the resonance frequency is still 503.3 MHz where the input impedance is R = 10 Ω. Again this shows that the impedance of the inductor cancels the impedance of the capacitor at resonance. In other words, the reactance, X_L is equal to the reactance, $X_{C,}$ at the resonance frequency.

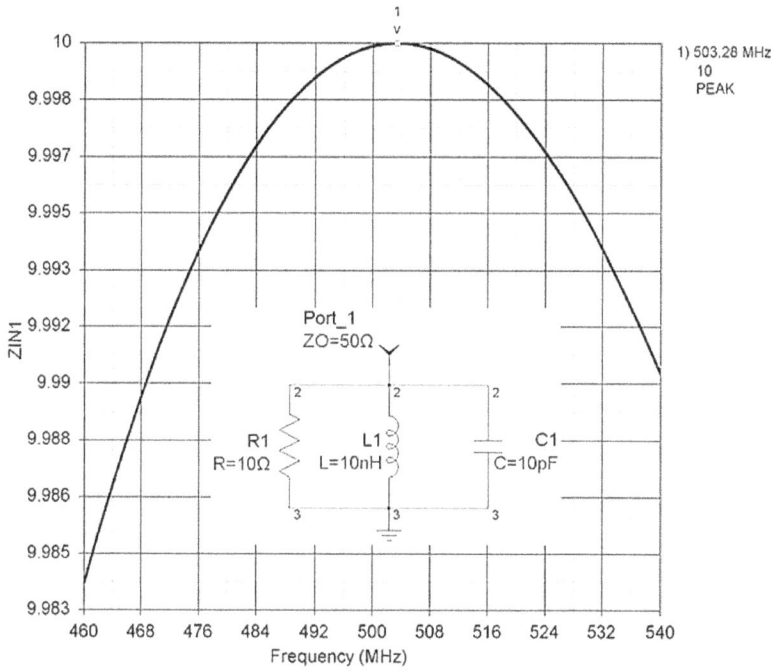

Figure 4-2 One-port parallel RLC resonant circuit and input impedance

The input admittance of the parallel resonant circuit is given by:

$$Y_{IN} = \frac{1}{R} + j\omega C - j\frac{1}{\omega L}$$

If the AC voltage across the parallel resonant circuit is V, then the complex power delivered to the resonator is:

$$P_{in} = \frac{|V|^2}{2} Y_{in} = \frac{|V|^2}{2}\left(\frac{1}{R} + j\omega C - j\frac{1}{\omega L}\right)$$

At resonance the reactive power of the inductor is equal to the reactive power of the capacitor. Therefore, the power delivered to the resonator is equal to the power dissipated in the resistor.

$$P_{in} = \frac{|V|^2}{2R}$$

The resonance frequency for the parallel resonant circuit as well as the series resonant circuit is obtained by setting,

$$\omega_0 C = \frac{1}{\omega_0 L} \text{ or:}$$

$$\omega_o = 2\pi f_o = \frac{1}{\sqrt{L\,C}}$$

Where, ω_0 is the angular frequency and f_0 is equal to the frequency in Hertz.

4.3 Resonant Circuit Loss

In Figure 4-1 and 4-2 the resistor, R1, represents the loss in the resonator. It includes both the losses in the capacitor as well as the inductor. The Q factor can be shown to be a ratio of the energy stored in the inductor and capacitor to the power dissipated in the resistor as a function of frequency.

For the series resonant circuit we have:

$$Q_u = \frac{X}{R} = \frac{\omega_o L}{R} = \frac{1}{\omega_o RC}$$

The Q factor of the parallel resonant circuit is simply the inverse of the series resonant circuit.

$$Q_u = \frac{R}{X} = \frac{R}{\omega_o L} = \omega_o RC$$

Notice that as the resistance increases in the series resonant circuit, the Q factor decreases. Conversely as the resistance increases in the parallel resonant circuit, the Q factor increases. The Q factor is a measure of loss in the resonant circuit. Thus a higher Q corresponds to lower loss and a lower Q corresponds to a higher loss. It is usually desirable to achieve high Q factors in a resonator as it will lead to lower losses in filters or lower phase noise in oscillators. Note that the resonator Q is defined as the unloaded Q of the resonator. This means that the resonator is not connected to any source or load impedance and as such is unloaded. The measurement of Q_u requires that the resonator be attached (coupled) to a signal source or load of some finite impedance. The equations would then have to be modified to include the source and load resistance. We might also surmise that any reactance associated with the source or load impedance may alter the resonant frequency of the resonator. This leads to two additional definitions of Q factor; namely the loaded Q and external Q.

4.4 Loaded Q and External Q

Example 4-3: Analyze the parallel resonator that is attached to a 50 Ohm source and load resistors.

Solution: To define the Q factor for the circuit requires that we include the source and load resistance which is 'loading' the resonator. This leads to the definition of the loaded Q, Q_L, for the parallel resonator as given in next equation .

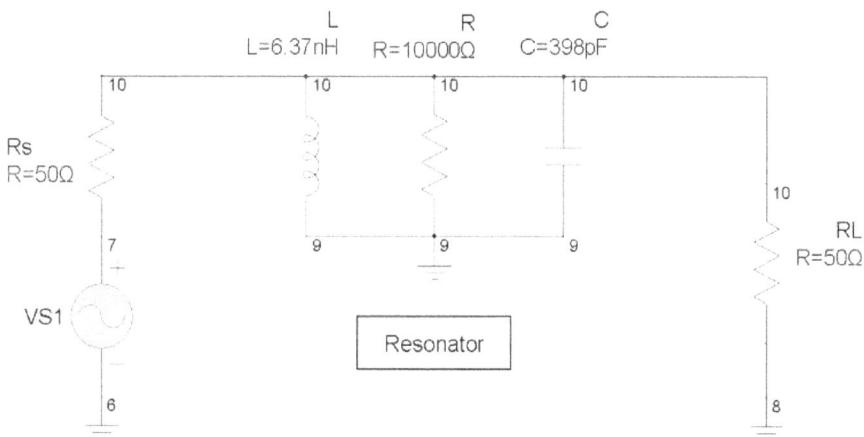

Figure 4-3 Parallel resonator with source and load impedance attached

$$Q_L = \frac{R_S + R + R_L}{\omega_o L}$$

Conversely we can define a Q factor in terms of only the external source and load resistance. This leads to the definition of the external Q, Q_E.

$$Q_E = \frac{R_S + R_L}{\omega_o L}$$

The three Q factors are related by the inverse relationship.

$$\frac{1}{Q_L} = \frac{1}{Q_E} + \frac{1}{Q_U}$$

At RF and microwave frequencies it is difficult to directly measure the Q_u of a resonator. We may be able to calculate the Q factor based on the physical properties of the individual inductors and capacitors as we seen in chapter 1. This is usually quite difficult and the Q factor is typically measured using a Vector Network Analyzer, VNA. Therefore, the measured Q factor is usually the loaded Q, Q_L. External Q is often used with oscillator circuits that are generating a signal. In this case the oscillator's load impedance is varied so that the external Q can be measured. The loaded Q of the network is then related to the fractional bandwidth.

$$Q_L = \frac{f_o}{BW}$$

Where, BW is the 3 dB bandwidth in Hertz and f_0 is equal to the resonant frequency in Hertz.

4.5 Lumped Element Parallel Resonator Design

Example 4-4: Design a lumped element parallel resonator at a frequency of 100 MHz. The resonator is intended to operate between a source resistance of 100 Ω and a load resistance of 400 Ohm.

Solution: Best accuracy would be obtained by using S parameter files or Modelithics models for the inductor and capacitor. However a good first order model can be obtained by using the Genesys inductor and capacitor models that include the component Q factor. These models save us the work of calculating the equivalent resistive part of the inductor and capacitor model. Use the Q factors shown in the schematic of Figure 4-4.

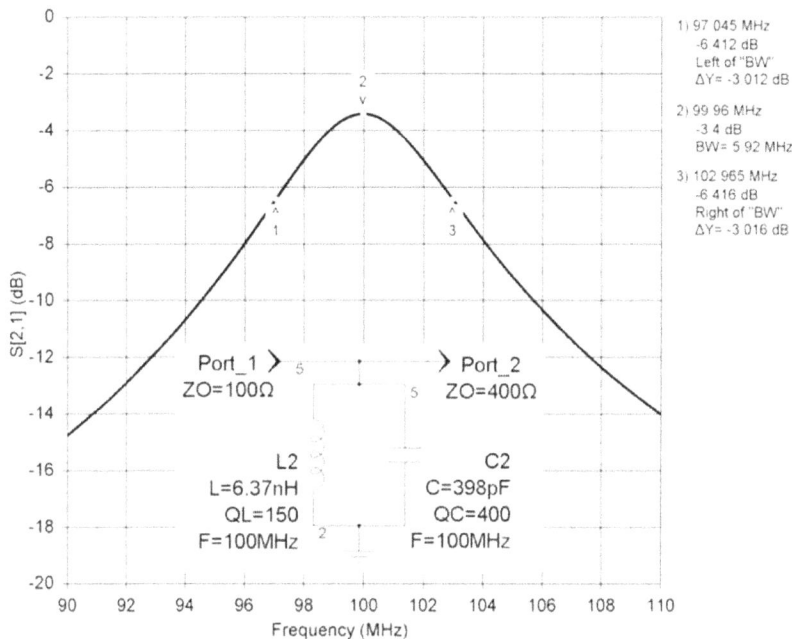

Figure 4-4 Resonator using inductor and capacitor with assigned Q values

Note that markers have been placed on the plot of the insertion loss, S21 that gives a direct readout of the 3 dB bandwidth. The loaded Q, Q_L can be calculated using:

$$Q_L = \frac{f_o}{BW} = \frac{99.955\,MHz}{5.923\,MHz} = 16.87$$

The designer must use caution when sweeping resonant circuits in Genesys. Particularly high Q band pass networks require a large number of discrete frequency steps in order to achieve the necessary resolution required to accurately measure the 3dB bandwidth. In this example the Linear Analysis is set up to sweep the circuit from 90 MHz to 110 MHz using 2000 points.

Place a marker anywhere on the trace and double click to open the Marker Properties window. Enter any name for the marker, select Bandwidth (Tracks Peak), and make sure that -3.01dB is entered as the relative offset. As Figure 4-4 shows the 3 dB bandwidth is automatically calculated as 5.923 MHz. The frequency peak and bandwidth label next to marker 2 is then used to calculate Q_L.

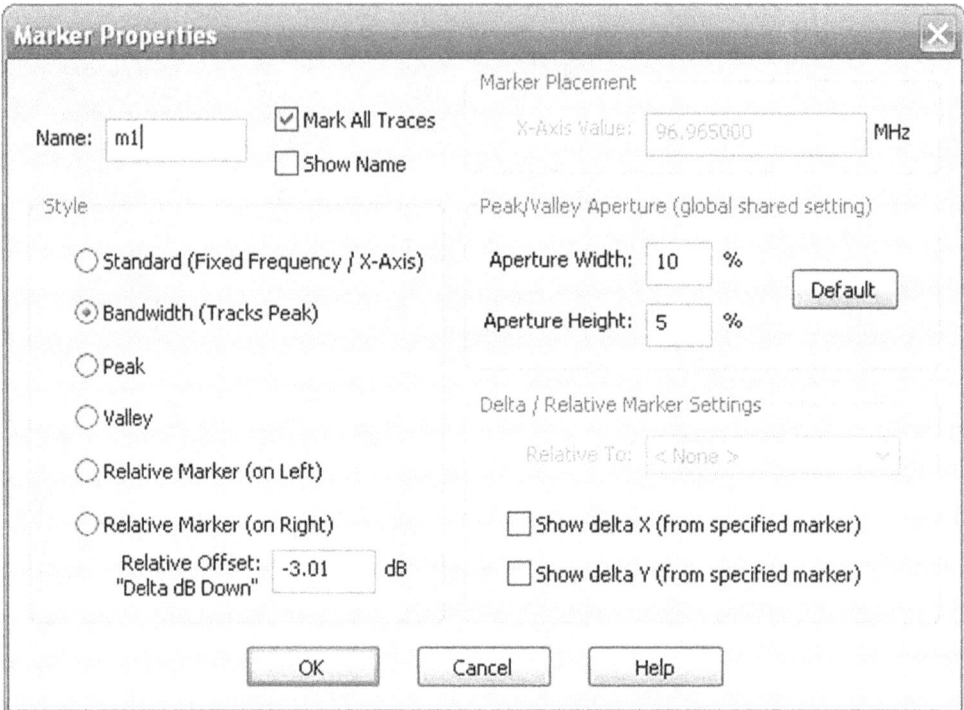

Figure 4-5 Bandwidth marker settings for measurement of 3dB bandwidth

4.6 Effect of Load Resistance on Bandwidth and Q_L

In RF circuits and systems the impedances encountered are often quite low, ranging from 1 Ω to 50 Ω. It may not be practical to have a source impedance of 100 Ω and a load impedance of 400 Ω.

Example 4-5: Using the parallel LC resonator of example 4-4, change the load resistance to 50 Ohm and re-examine the circuit's 3 dB bandwidth and Q_L.

Solution: The 3 dB bandwidth is now 12.926 MHz resulting in a loaded Q factor of 7.73. The loaded Q factor has decreased by nearly half of the original value. We have increased the bandwidth or de-Q'd the resonator.

This can also be thought of as tighter coupling of the resonator to the load.

$$Q_L = \frac{f_o}{BW} = \frac{99.965\,MHz}{12.926\,MHz} = 7.73$$

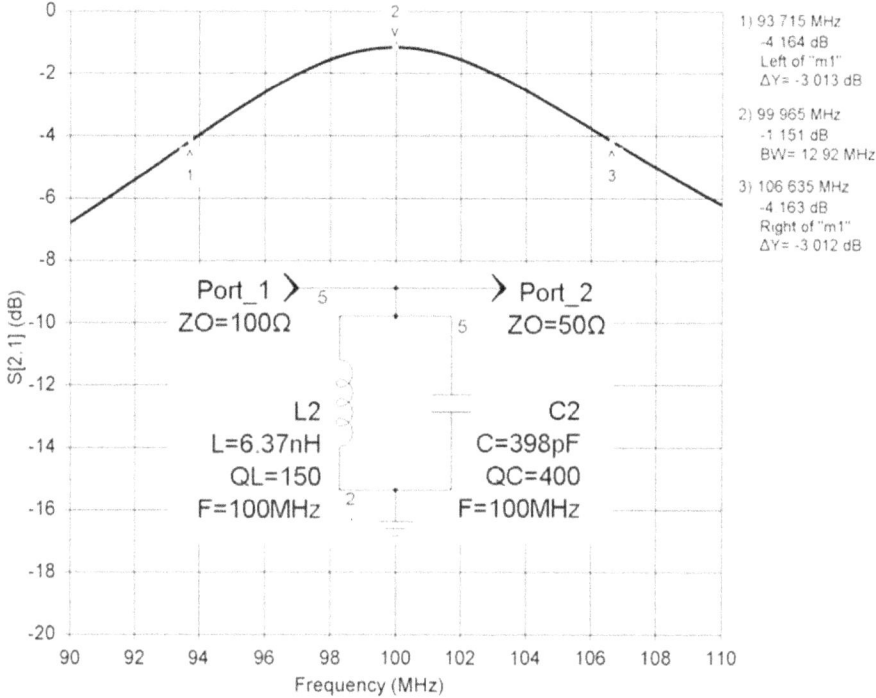

Figure 4-6 A parallel resonance circuit showing the 3dB bandwidth

4.7 Lumped Element Resonator Decoupling

To maintain the high Q of the resonator when attached to a load, such as 50 Ohm, it is necessary to transform the low impedance to high impedance presented to the load. The 50 Ohm impedance can be transformed to the higher impedance of the parallel resonator thereby resulting in less loading of the resonator impedance. This is referred to as loosely coupling the resonator to the load. The tapped-capacitor and tapped-inductor networks can be used to accomplish this Q transformation in lumped element circuits.

4.8 Tapped Capacitor Resonator

Example 4-6: Consider rearranging the parallel LC network of Example of 4-5 by replacing the capacitor with tapped capacitor C1 and C2. Measure the circuit's 3 dB bandwidth and loaded Q_L of the tapped capacitor network.

Solution: The new capacitor values for C1 and C2 can be found by the simultaneous solution of the following equations.

$$C_T = \frac{C1 \cdot C2}{C1 + C2}$$

$$R_{L1} = R_L \left(1 + \frac{C1}{C2}\right)^2$$

R_{L1} is the higher, transformed, load resistance. In this example substitute R_{L1} = 400 Ω, the original load resistance value. C_T is simply the original capacitance of 398 pf. The capacitor values are found to be: C1 = 1126.23 pF and C2 = 616.1 pF. The new resonator circuit is shown in Figure 4-7. Sweeping the circuit we see that the response has returned to the original performance of Figure 4-6. The 3 dB bandwidth has returned to 5.923 MHz making the Q_L equal to:

$$Q_L = \frac{99.925}{5.923} = 16.87$$

The 50 Ω load resistor has been successfully decoupled from the resonator. The tapped capacitor and inductor resonators are popular methods of decoupling RF and lower microwave frequency resonators. It is frequently seen in RF oscillator topologies such as the Colpitts oscillator in the VHF frequency range.

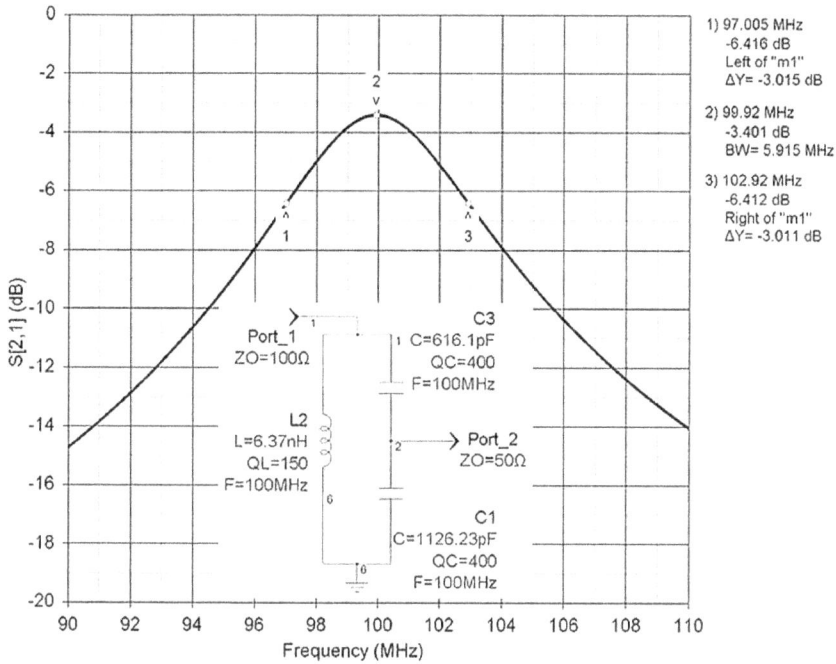

Figure 4-7 Parallel LC resonator using a tapped capacitor and response

4.9 Tapped Inductor Resonator

Example 4-7: Design a tapped inductor network to decouple the 50 Ohm source impedance from loading the resonator.

Solution: Replace the 100 Ω source impedance with a 50 Ω source and use a tapped inductor network to transform the new 50 Ω source to 100 Ω. Modify the circuit to split the 6.37 nH inductor, L_T, into two series inductors, L1 and L2. The inductor values can then be calculated by solving the following equation set simultaneously. R_{S1} is the higher, transformed, source resistance. In this example substitute R_{S1} = 100 Ω,

$$Rs1 = Rs\left(\frac{L_T}{L_1}\right)^2$$

$$L_T = L_1 + L_2$$

Solving the equation set results in values of L1=4.5 nH and L2=1.87 nH. The resulting schematic and response is shown in Figure 4-8. The new response is identical to the plot of Figure 4-5. Therefore we now have a source and load resistance of 50 Ω and have not reduced the Q of the resonator from what we had with the original source resistance of 100 Ohm and a load resistance of 400 Ω.

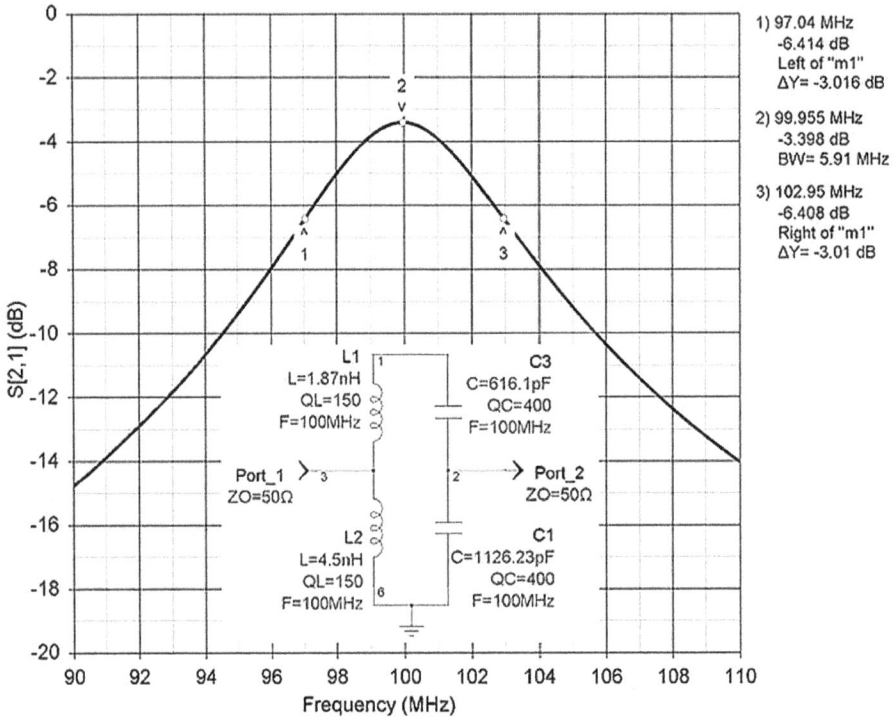

Figure 4-8 Tapped-inductor added to the parallel resonant circuit

4.10 Genesys Model of the Microstrip Resonator

The half wave open circuit microstrip resonator is modeled in Genesys as shown in Figure 4-9. Note that the source and load impedance has been increased to 5000 Ω to avoid loading the impedance of the parallel resonant circuit.

Perform a linear sweep of the resonator using 4001 points from 4500 MHz to 5400 MHz.

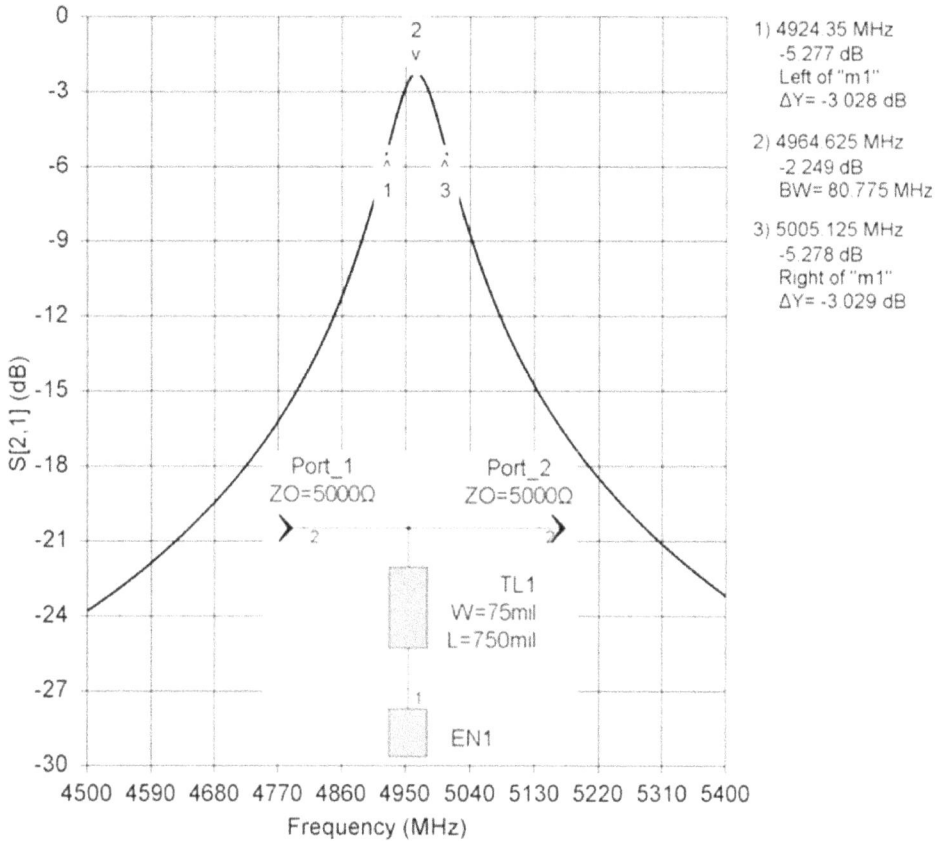

Figure 4-9 Half-wave open circuit microstrip resonator

Using the techniques of section 4.2.3 and Equation (4-10), the 3dB bandwidth is measured to determine the loaded Q of the resonator.

$$Q_L = \frac{4964.625}{80.775} = 61.5$$

The insertion loss at the resonant frequency can be used to relate the Q_L to the Q_u as shown by Equation (4-25). The Q_u as simulated by Genesys is within 20% of the value calculated using the substrate physical and electrical characteristics.

$$InsertionLoss(dB) = 20\log\frac{Q_u}{Q_u - Q_L}$$

$$Q_u = \frac{\left(10^{IL(dB)/20}\right)Q_L}{\left(10^{IL(dB)/20}\right)-1} = \frac{\left(10^{2.249/20}\right)(61.5)}{\left(10^{2.249/20}\right)-1} = 269.5$$

It is also interesting to note that the Q_L of the resonator is related to the group delay through the two-port network.

$$Q_L = 2\pi f\left(\frac{t_d}{2}\right)$$

where:

t_d is the group delay in seconds
f is the frequency in Hertz

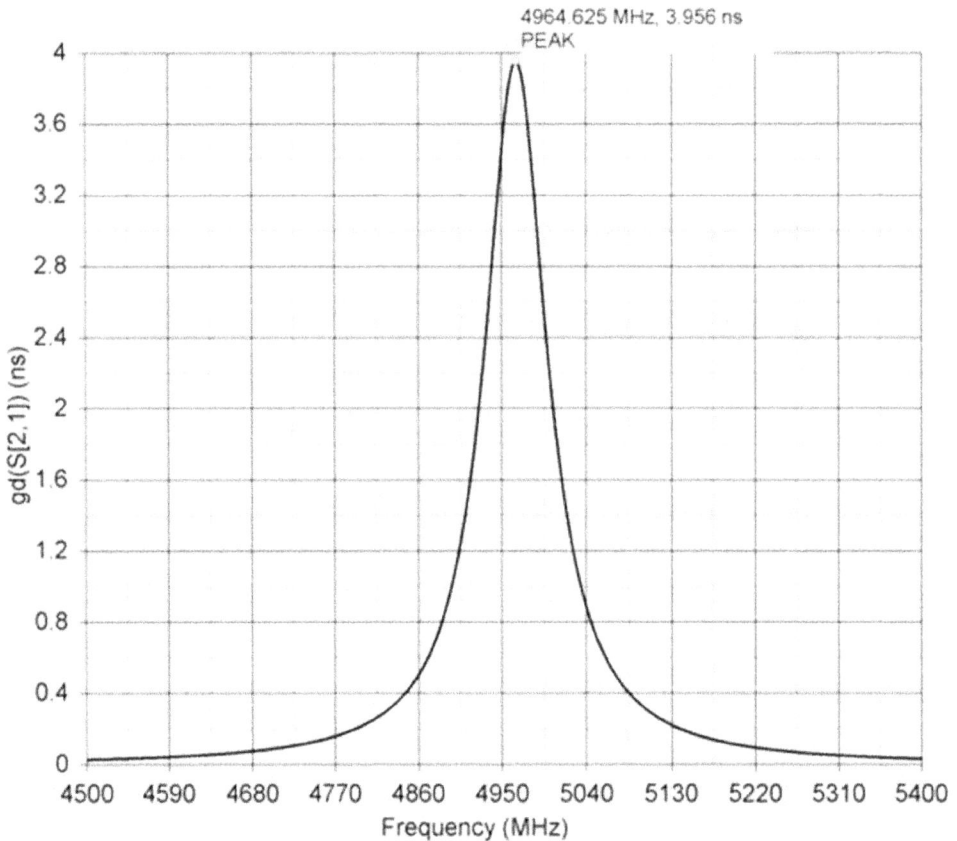

Figure 4-10: Group delay across the half-wave open circuit microstrip resonator

4.11 Resonator Series Reactance Coupling

Example 4-8: A 5 GHz half wavelength open-circuited microstrip resonator is capacitively coupled on Roger's RO3003 dielectric material. Find the value of the capacitors for the under-coupled, critically coupled, and over-coupled cases.

Solution: Figure 4-11 shows the capacitor values for the 3 cases.

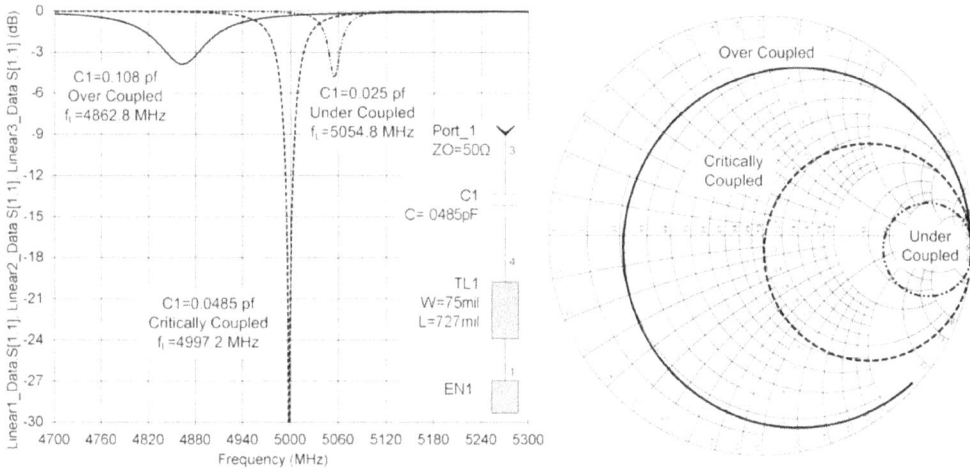

Figure 4-11 Capacitive coupled half-wave microstrip resonator

Example 4-9A: Use the analytical technique to measure the Qo and Q_L of the critically-coupled half wave microstrip resonator.

Solution: The equations in Table 4-1 shows how to calculate the Q_0 factor.

```
angle=175.348
circlediameter=7440.36
couplingLoss=0.02
circleangle=-angle/2
GammaD=0.999
GammaL=0.574
fL=5054.8
f1=5066.4
f2=5043.2
d=GammaD-GammaL
k=d/(2-d)
QL=fL/abs(f1-f2)
Qo=QL*(1+k)
```

Table 4-1: Calculation of Q_0 and Q_L factors

The result of calculation is shown in Table 4-2.

Index	QL	Qo
1	217.879	276.672

Table 4-2: Q_L and Q_0 values

4-12 Filter Design at RF and Microwave Frequency

We have seen that it is possible to change the shape of the frequency response of a parallel resonant circuit by choosing different source and load impedance values. Likewise multiple resonators can be coupled to one another and to the source and load to achieve various frequency shaping responses. These frequency shaped networks are referred to as filters.

Fortunately these tables have been built into many filter synthesis software applications that are readily available. In this book we will examine the filter synthesis tool that is built into the Genesys software. We will work through two practical filter examples, one low pass and one high pass using the Genesys filter synthesis tool.

4.13 Low Pass Filter Design

Example 4-10: Design a 415 MHz low pass filter with the following specifications:

(a) Select a Chebyshev Response with 0.1 dB pass band ripple.

(b) Set the passband cutoff frequency (not the -3 dB frequency) at 160 MHz

(c) The reject requirement is at least -40 dB rejection at 435 MHz.

The transmitter and receiver antennas are on the same physical support boom so there is limited isolation between the transmitter and receiver. Even though the signals are at different frequencies, the broadband noise amplified by the power amplifier at 435 MHz will be received by the UHF antenna and sent to the sensitive receiver. Because the receiver is trying to detect very low signal levels, the received noise from the amplifier will interfere or 'de-sense' the received signals. Therefore it is necessary to design a 145 MHz Low Pass filter for this satellite link system. The specifications chosen for the filter design are selected as:

Solution: Use the Passive Filter Synthesis utility to design the Low Pass filter. Select a Low Pass filter of the Chebyshev type. On the Topology tab select a Lowpass filter type with a Chebyshev shape. Select the minimum capacitor subtype. On the Settings tab enter the cutoff frequency of 160 MHz and pass band ripple of 0.1 dB. Also set the cutoff frequency attenuation at 0.1 dB. The Filter Settings Tab is a great place to perform "what-if" analysis. The pass band ripple, filter order, cutoff frequency, and attenuation at cutoff can all be varied while observing their impact on the filter's characteristic. For the design example enter the parameters as shown in Fig. 4-12. The required filter order can be determined by increasing the order until the specification of -40 dBc attenuation at 435 MHz is achieved. As the filter response curve in Fig.4-12 shows, this Low Pass filter must be of fifth order to achieve the required attenuation. Along with the resulting attenuation and return loss the synthesis program creates the filter schematic with the synthesized component values.

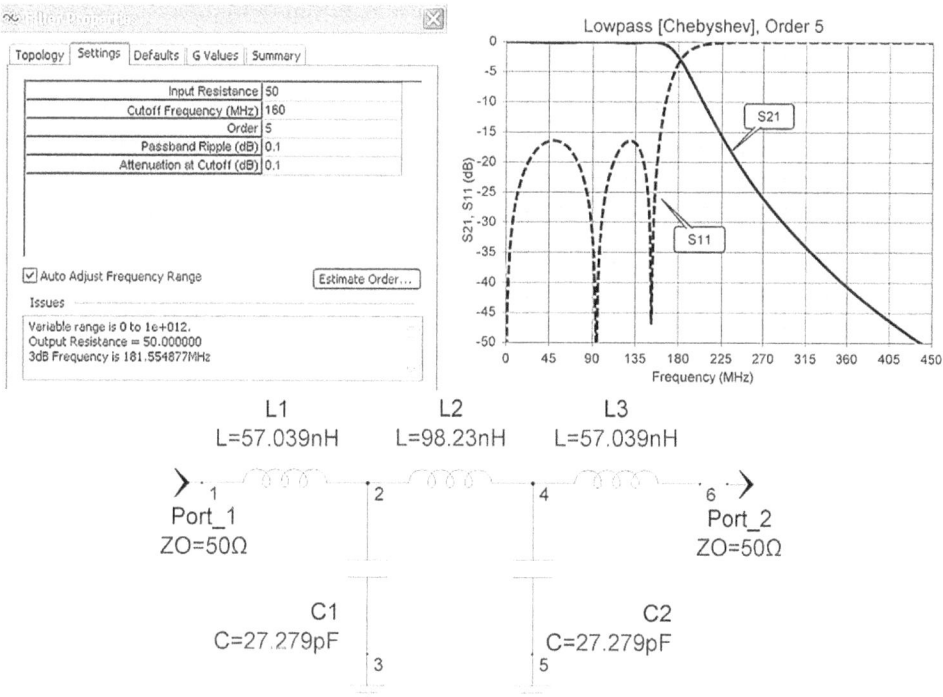

Figure 4-12 Passive filter synthesis utility: topology and settings tab

4.14 Physical Model of the Low Pass Filter

In this section we want to design a physical low pass filter and display S11 and S21. The synthesized filter is an ideal design in the sense that ideal components have been used. To obtain a 'real-world' simulation of the filter we need to use component models that have finite Q and parasitics such as multilayer chip capacitors for shunt capacitors. For power handling capability, choose the 700 series chip capacitors from ATC Corporation. We will use measured S parameter files to model the shunt capacitors. ATC Corporation has a useful application for selection of chip capacitors called ATC Tech Select. Because the filter is passing relatively high power (20 W), we cannot use small surface mount style chip inductors. Instead we use air wound coils to realize the series inductors. The inductors will be realized with AWG#16 wire nickel-tin plated copper wire. The wire has a diameter of 0.05 inches or 50 mils. They will be wound on a 0.141 inch diameter form. Use the techniques covered in Chapter one with the Air Wound inductor model in Genesys. The filter model is then reconstructed using the S parameter files for the shunt capacitors and the physical inductor models for the series inductors. Make sure to model the substrate and the interconnecting printed circuit board traces as microstrip lines. Also model the ground connection of the shunt capacitors as a microstrip via hole. Although these PCB parasitic effects are normally more pronounced at above 2 GHz.

It often surprising the effect that these parasitics have at lower frequencies. The final filter response and model is shown in Figure 4-14. The response shows that the attenuation specification has been achieved. Because the circuit has physical models replacing the ideal lumped element components, the engineer can be confident that the filter can be assembled and will achieve the responce.

Figure 4-13 is a photo of the prototype low pass filter circuit with SMA coaxial connectors attached to the circuit board.

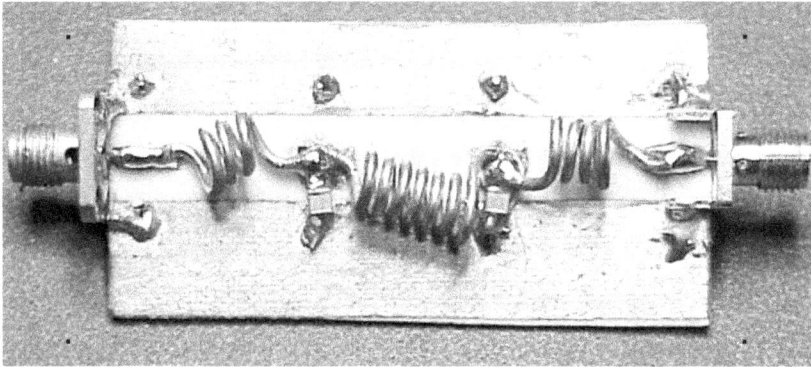

Figure 4-13 Physical prototype of the 146 MHz low pass filter

The schematic and response of the filter are shown in Figure 4-14.

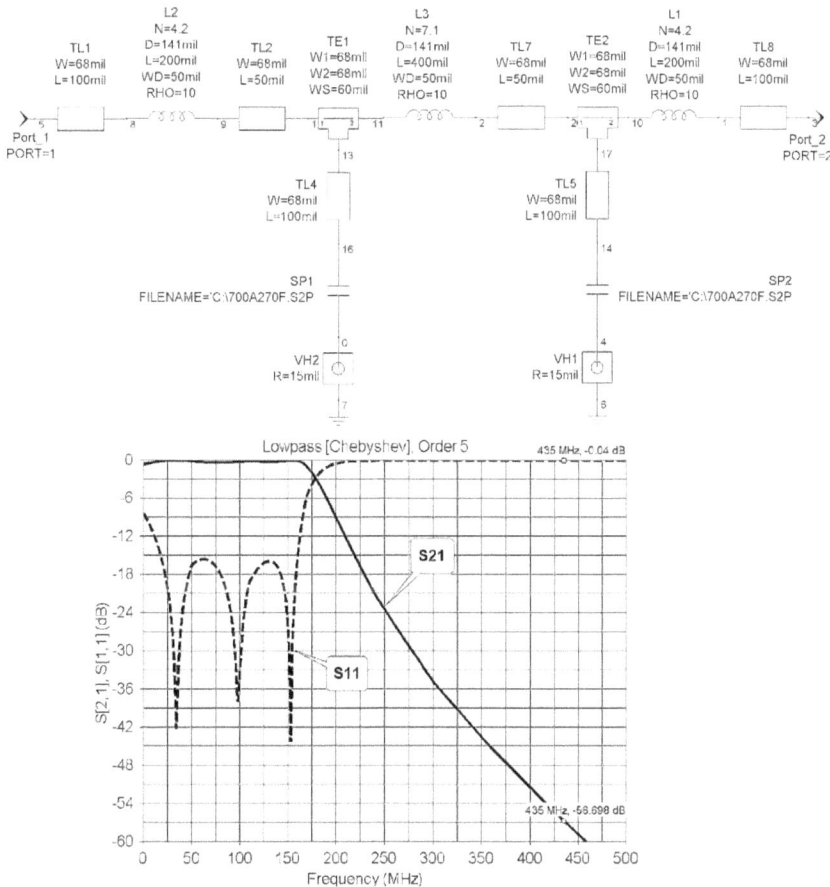

Figure 4-14 Low pass filter model with physical elements

4.15 High Pass Filter Design Example

Example 4-11: Design a high pass filter that passes frequencies in the 420 MHz to 450 MHz range. This filter could be placed in front of the preamp used in the downlink of the satellite system. This would help to keep out any of the transmit energy or noise power in the 146 MHz transmit frequency range. The High Pass Filter specifications are:

- The pass band cutoff frequency (not the -3 dB frequency) is 420 MHz.
- The filter has a Chebyshev response with 0.1 dB pass band ripple.
- The reject requirement is at least -60 dB rejection at 146 MHz.

Solution: Using the Passive Filter Synthesis tool in Genesys, vary the filter order until sufficient attenuation is achieved at 146 MHz. It is always a good practice to design for some additional rejection (margin) that exceeds the minimum requirement. The Filter synthesis session recommends a filter of 7th order.

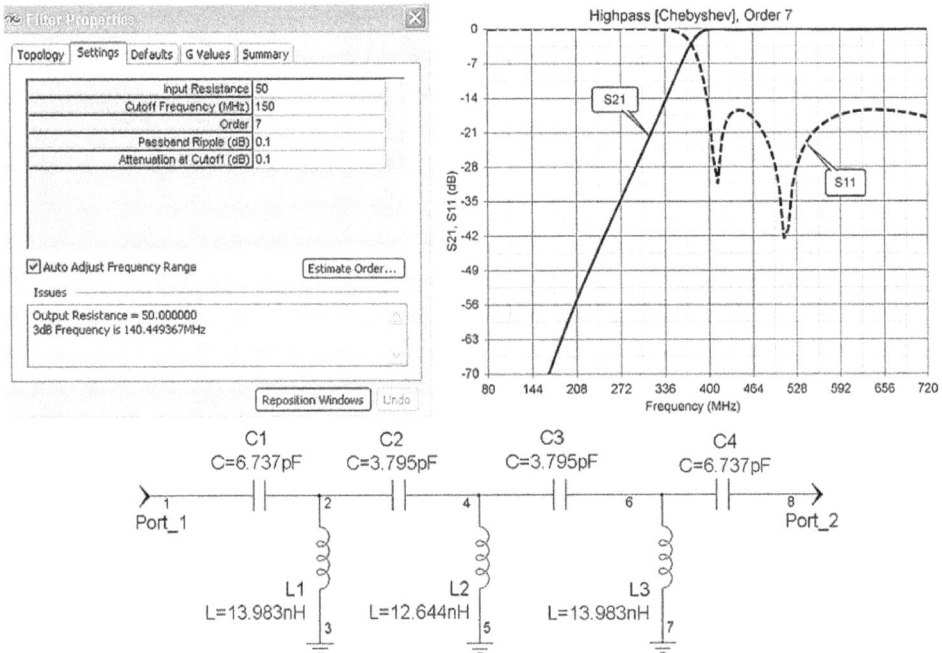

Figure 4-15 Passive filter synthesis model of the high pass filter

4.16 Physical Model of the High Pass Filter

Using the same techniques as described for the Low Pass Filter design we can proceed with the High Pass Filter realization. The Passive Filter Synthesis application calculated series capacitance values of 6.7 pF and 3.8 pF. Looking through the available ATC 700 series chip capacitors, the nearest values are 6.8 pF and 3.9 pF capacitors. We will select the S parameters files for these chip capacitors to use in the final filter model. The shunt inductors are realized using the Air Wound inductor model. Figure 4-16 shows the final circuit model and filter response. From the response we can see that the required rejection specification of S21 < -60 dBc has been maintained. Figure 4-16 shows the completed assembly of the High Pass Filter on a printed circuit board with coaxial SMA connectors.

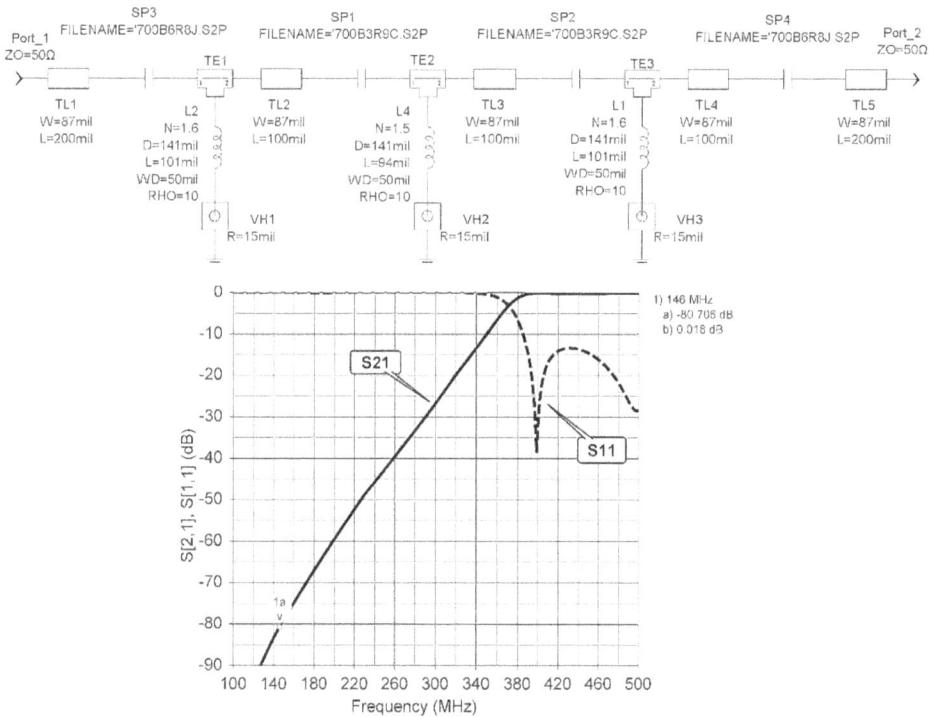

Figure 4-16 High pass filter model with physical elements

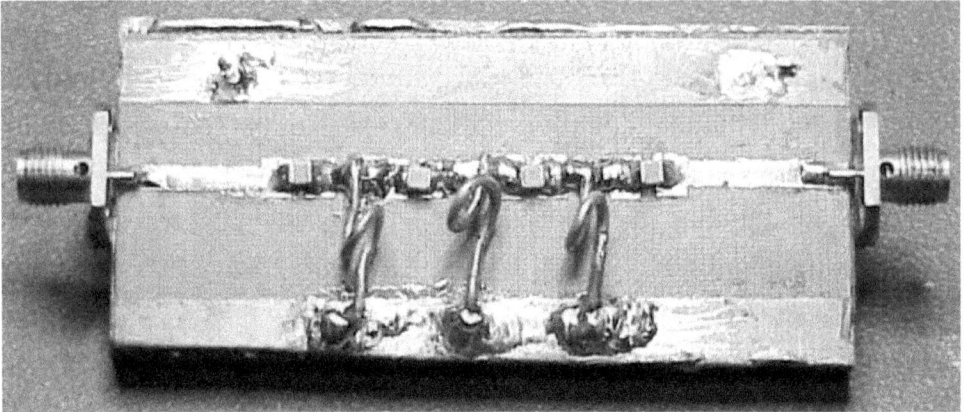

Figure 4-17 Physical prototype of the 420 MHz high pass filter

4.17 Microstrip Stepped Impedance Low Pass Filter

In the microwave frequency region filters can be designed using distributed transmission lines. Series inductors and shunt capacitors can be realized with microstrip transmission lines. In the next section we will explore the conversion of a lumped element low pass filter to a design that is realized entirely in microstrip.

4.18 Lumped to Distributed Element Conversion

Example 4-12: Design a lumped element 2.2 GHz Chebyshev response low pass filter having 0.1 dB passband ripple and -40 dB rejection at 6 GHz. This low pass filter has a 3 dB bandwidth of approximately 2470 MHz.

Solution: Following the procedure discussed in Example 4.8-1, first we design the low pass filter using the lumped elements. It is always a good practice to design for some additional rejection (margin) that exceeds the minimum requirement. The Filter synthesis session results in the lumped low pass filter of Figure 4-18.

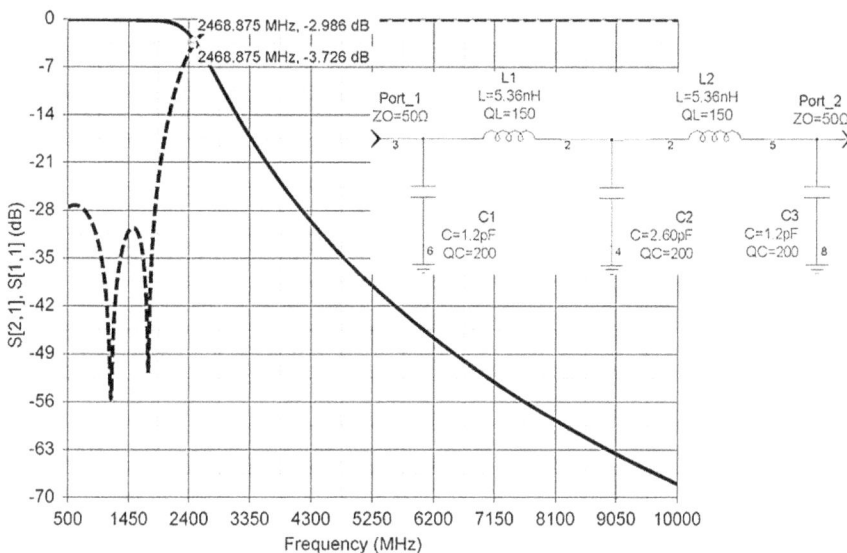

Figure 4-18 Lumped element 2.2 GHz low pass filter and frequency
response

Example 4-13: Using the Transformation Assistance in Genesys, transform
the lumped element low pass filter of Example 4.9-1A into distributed element
Filter. Simulate the schematic and display the response.

Solution: The series inductors will be realized as 80 Ω transmission lines of
sufficient length to act as a 5.36 nH inductor. The TLINE Transmission
Line synthesis program is used to calculate the microstrip line width for an
80 Ω transmission line on the Rogers 0.025 in. RO3010 material. As Figure
4-19 shows the 80 Ω transmission line has a line width of 6.26 mils with an
effective dielectric constant of 6.08. To realize the required inductance
value a specific length of 80 Ω transmission line is required.

Equation Editor of Figure 4-20 is used to calculate the length of microstrip
line. The length of line required for a 5.36 nH inductor is then found to be
321 mils. The shunt capacitors will be realized as 20 Ω transmission lines.
Using TLINE the 20 Ω line width is calculated to be 102.8 mils with an
effective dielectric constant of 8.00. Equation Editor of Figure 4-21shjows
that the length for the 2.6 pF shunt capacitor is 217 mils. The line length for
the 1.2 pF capacitors is then found to be 100 mils.

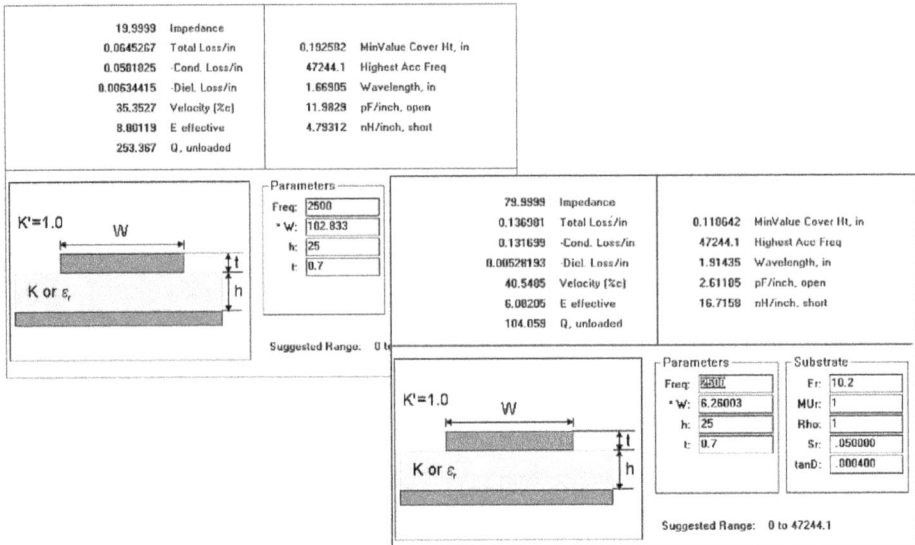

Figure 4-19 TLINE calculations for 20 Ω and 80 Ω microstrip lines

```
F1=2.5e9
erEFF1=6.08
L=5.36e-9
Z1=80
Lambda1=3e8/F1
LambdaG1=Lambda1/sqrt(erEFF1)/.0254
Length1=F1*LambdaG1*L/Z1
```

Figure 4-20 Calculating the length of inductive microstrip line

```
F=2.5e9
Z=20
erEFF=8.00
C=2.6e-12
LambdaG=Lambda/sqrt(erEFF)/.0254
Length=LambdaG*F*Z*C
```

Figure 4-21 Calculating the length of capacitive microstrip line

The length of the microstrip lines are 0.321and 0.217 inches, respectively. Adding a short 50 Ω section to the input and output, the initial low pass filter schematic and response is shown in Figure 4-21.

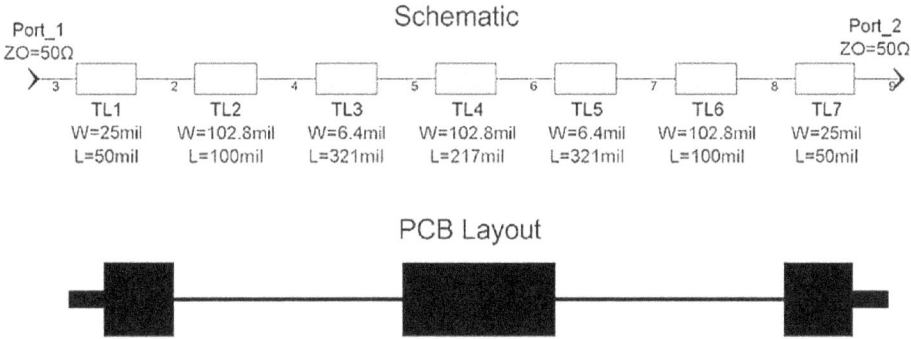

Figure 4-22 Initial schematic and PCB layout of the low pass filter

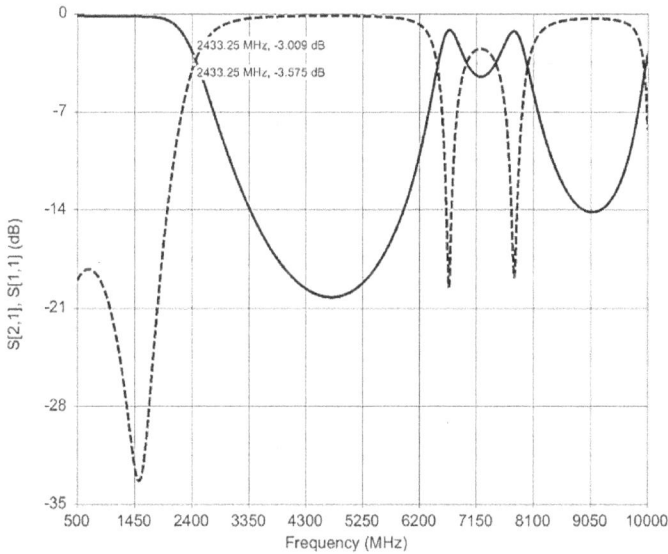

Figure 4-23 Initial Response of the microstrip Low Pass Filter

Examine the PCB layout of the Low Pass Filter of Figure 4-22. Note the change in geometry as the impedance transitions from 50 Ω to 20 Ω and from 20 Ω to 80 Ω. These abrupt changes in geometry are known as discontinuities. Genesys has several model elements that can help to account for the effects of discontinuities. These include: T-junctions, cross junctions, open circuit end effects, coupling gaps, and bends.

Example 4-14: Add microstrip Step elemenets between microstrip lines of different widths and re-simulate to see the slight shift in the 3 dB points. The microstrip substrate is Rogers's 6010 material with a 0.025 dielectric thickness.

Solution: A Microstrip Step element can be placed between series lines of abruptly changing geometry to account for the step discontinuity. Place the Microstrip Step element at each impedance transition in the filter. Make sure that the narrow side and wide side are directed appropriately. The Microstrip Step element will automatically use the adjacent width in its calculation. Figure 4-24 shows the low pass filter schematic with the step elements added between transmission line sections.

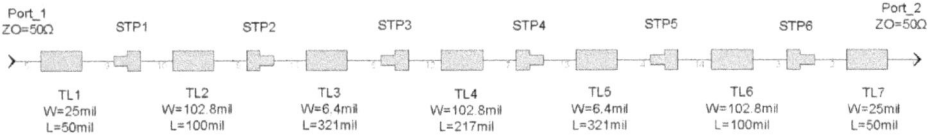

Figure 4-24 Stepped impedance filter with added "step" elements

A comparison of the initial low pass filter model and the modified model is shown in Figure 4-25. As the Figure shows there is a slight difference in the filter insertion loss, S21, particularly as the frequency increases from the cutoff at 2400 MHz.

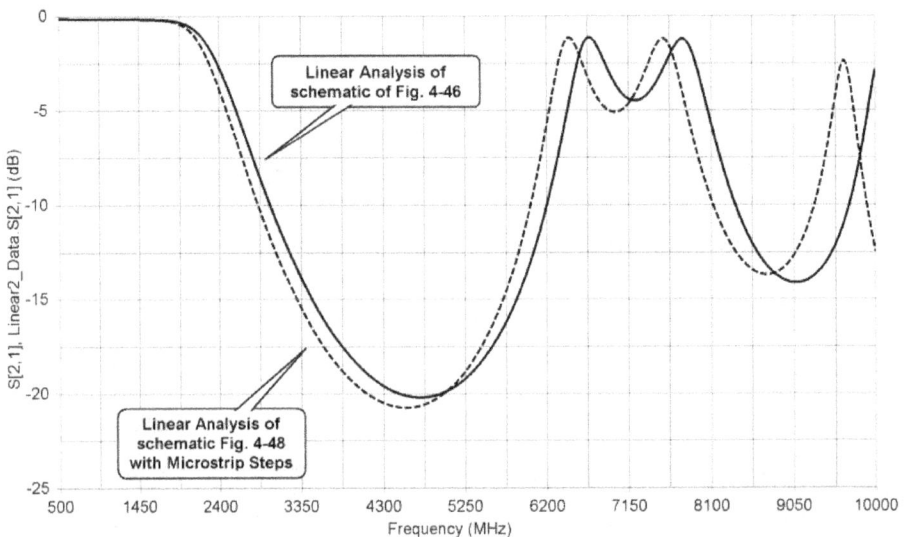

Figure 4-25 Filter cutoff frequency shift due to step discontinuities

4.19 Microstrip Coupled Line Filter Design

The edge coupled microstrip line is very popular in the design of band pass filters. A cascade of half-wave resonators in which quarter wave sections are parallel edge coupled lines, are very useful for realizing narrow band, band pass filters. This type of filter can typically achieve $\leq$ 15% fractional bandwidths [4].

Example 4-15: Design a band pass filter at 10.5 GHz. The filter is designed on RO3010 substrate (ε_r = 10.2) with a dielectric thickness of 0.025 inches. The filter should have a pass band of 9.98 – 11.03 GHz. As a design goal the filter should achieve at least 20 dB rejection at 9.65 GHz. In other words the filter is required to have > 20 dB rejection at 330 MHz below the lower passband frequency.

Solution: The Microwave Filter synthesis utility in Genesys is used to design the filter network. Figure 4-26 shows the entry of the design parameters into the Topology, Settings, and Options tabs.

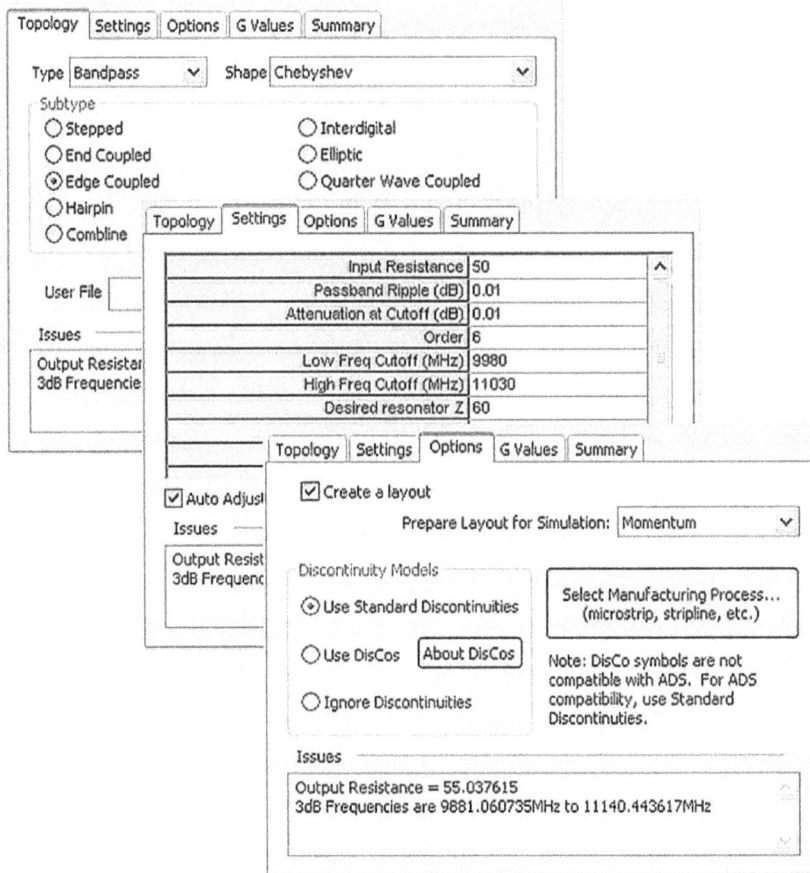

Figure 4-26 Bandpass filter synthesis selections

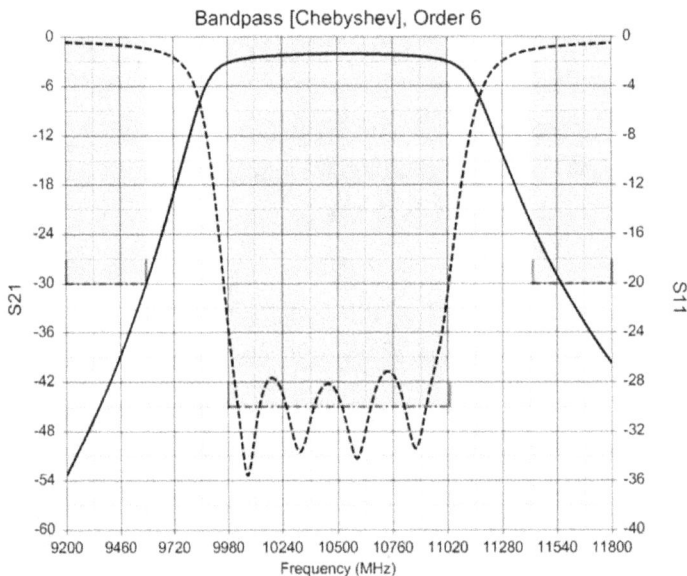

Figure 4-27 Bandpass filter response

On the Topology tab set the design for an edge coupled band pass filter with a Chebyshev shape. On the Settings tab, enter the pass band frequencies per the filter specification. Vary the order until the desired filter rejection is achieved. The Microwave Filter synthesis program shows that a 6th order filter should meet the rejection specifications. On the Options tab, select standard discontinuities and check the 'Create a Layout'. The synthesized filter schematic is shown in Figure 4-28.

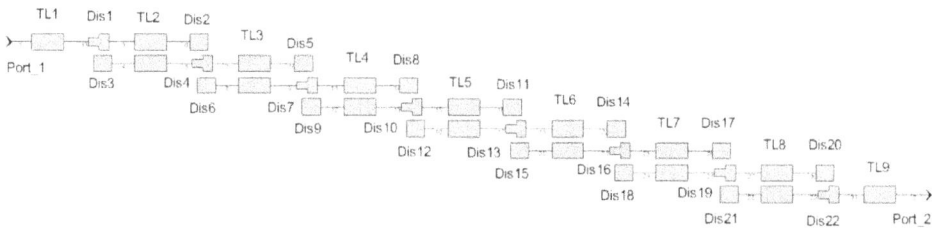

Figure 4-28 Parallel line edge coupled microstrip bandpass filter schematic

References and Further Reading

[1]   Ali Behagi and Stephen D. Turner, *Microwave and RF Engineering,* A Simulation Approach with Keysight Genesys Software, BT Microwave LLC, March, 2015

[2]   RF Circuit Design, Second Edition, Christopher Bowick, Elsevier 2008

[3]   Keysight Technologies, *Genesys 2015.08, Users Guide,* www.keysight.com

[4]   Foundations for Microstrip Circuit Design, T.C. Edwards, John Wiley & Sons, New York, 1981

[5]   Q Factor, Darko Kajfez, Vector Fields, Oxford Mississippi, 1994

[6]   Q Factor Measurement with Network Analyzer, Darko Kajfez and
        Eugene Hwan, IEEE Transactions on Microwave Theory and
        Techniques, Vol. MTT-32, No. 7, July 1984.

[7]   High Frequency Techniques, Joseph F. White, John Wiley & Sons,
        Inc., 2005

[8]   Soft Substrates Conquer Hard Designs, James D. Woermbke,
        Microwaves, January 1982.

[9]   Principles of Microstrip Design, Alam Tam, RF Design, June 1988.

[10]   David M. Pozar, *Microwave Engineering*, Fourth Edition, John
        Wiley & Sons, New York, 2012

Problems

4-1.   Consider the one port resonator that is represented as a series RLC
        circuit as shown. Analyze the circuit, with R = 5 Ω, L = 5 nH, and C
        = 5 pF. Plot the magnitude of the resonator input impedance and
        measure the resonance frequency.

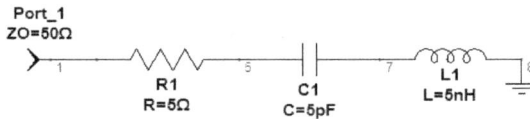

4-2.   Consider the one port resonator that is represented as a parallel RLC
        circuit as shown. Analyze the circuit, with R = 500 Ω, L = 50 nH,
        and C = 50 pF. Plot the magnitude of the resonator input impedance
        and measure the resonance frequency.

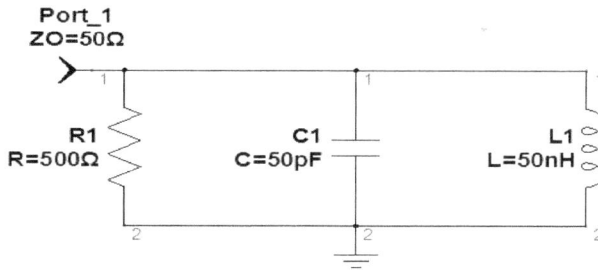

4-3. Design a Butterworth lowpass filter having a passband of 2 GHz with an attenuation 20 dB at 4 GHz. Plot the insertion loss versus frequency from 0 to 5 GHz. The system impedance is 50 Ω.

4-4. Design a 5th order Chebyshev highpass filter having 0.2 dB equal ripples in the passband and cutoff frequency of 2 GHz. The system impedance is 75 Ω. Plot the insertion loss versus frequency from 0 to 5 GHz.

4-5. In a full duplex communication link, the uplink signal is around 200 MHz while the downlink is at 500 MHz. A 25 Watt power amplifier is used on the uplink with 20 dB gain.

(a) Design a low pass filter on the uplink to pass the 200 MHz uplink signal while rejecting any noise power in the 500 MHz band. Design the passband cutoff frequency is at 220 MHz, therefore, the filter should have a Chebyshev Response with 0.1 dB pass band ripple. The reject requirement is at least -40 dB rejections at 500 MHz.

(b) Design a High Pass Filter that passes frequencies in the 480 MHz to 520 MHz range. The High Pass Filter specifications are: The passband cutoff frequency is 480 MHz, therefore, the filter should have a Chebyshev response with 0.1 dB pass band ripple. The reject requirement is at least -60 dB losses at 190 MHz.

4-6. Design a 75 Ω transmission line of sufficient length to act as a 10 nH inductor. Use the TLINE Transmission Line Synthesis Program to calculate the microstrip line width on the Rogers 0.025 inch RO3010 material.

4-7.   Using the microwave filter synthesis tool, design a stepped impedance low pass filter on RO3003 material that is 0.010 inches thick. Use a Chebyshev response with a 0.01dB ripple and a cutoff frequency of 4 GHz. Determine the worst case in band return loss and the rejection at 6 GHz.

4-8.   For the filter design of Problem 4-7 create an EM simulation using Momentum. Compare the EM simulation to a linear simulation. Comment on the rejection comparison at 6 GHz.

4-9.   For the filter design of Problem 4-7 determine the frequencies at which reentrant modes exist up through 20 GHz.

4-10.   Design a half wave microstrip resonator at 10 GHz using RO3003 substrate that is 0.020 thick. Initially design the resonator with a 50 W line impedance. Select a coupling capacitor to critically couple the resonator to the 50 Ω source. Then determine the resonator line impedance that results in the highest unloaded Q_o.

4-11.   Use the microwave filter synthesis tool to design a parallel edge coupled filter on RO3003 substrate that is 0.010 thick. Use a Chebyshev characteristic with 0.10 dB ripple. Design the passband to cover 10.7 GHz to 12.2 GHz. Determine the filter order required to achieve 30 dB rejection at 9 GHz.

4-12.   For the filter design of Problem 4-10, create an EM simulation using Momentum. Compare the linear and EM simulations. Determine the minimum box width (EM Box height, Y) that creates a box resonance frequency. What is the frequency of the box resonance?

Chapter 5

Power Transfer and Impedance Matching

5.1 Introduction

Impedance matching is an integral part of RF and microwave circuit and system design. It is necessary for the efficient transfer of power from a source to the load. For example, in microwave amplifier design, the need for impedance matching arises when the amplifier must be properly terminated at both terminals in order to deliver maximum power from the source to the load. In narrowband applications impedance matching can be achieved, at a single frequency, with a lossless two-element network, known as L-network. In this chapter the basics of power transfer and the conditions for maximum power transfer are presented. The mathematical equations for the design of discrete L-networks are derived, using Matlab compatible expressions.

5.2 Maximum Power Transfer Conditions

For the network of Figure 5-1 a voltage source, V_S, and the series impedance $Z_S = R_S + jX_S$ are connected to a network, having the input impedance $Z_{IN} = R_{IN} + jX_{IN}$, the power transferred to the network is given by Equation (5-1).

$$P_{Network} = \frac{\text{Re}\left[V_{IN} I_{IN}^*\right]}{2}$$

In this equation Re denotes the real part and the symbol * denotes the conjugate value.

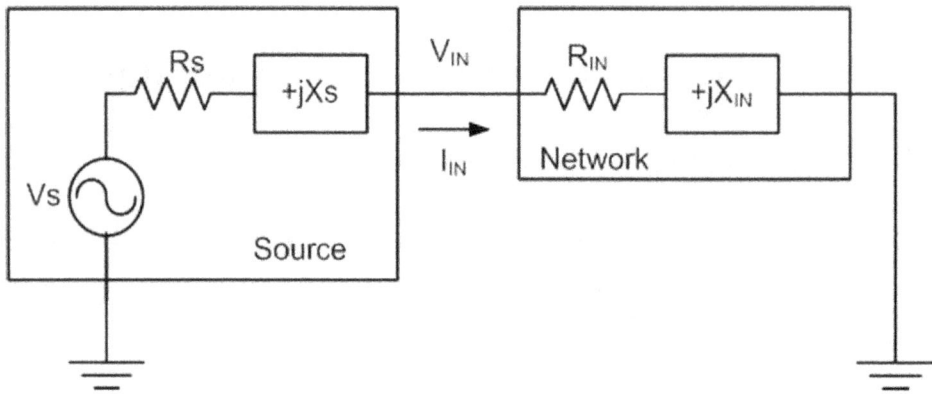

Figure 5-1 Voltage source connected to complex load impedance

The condition for maximum power transfer is that the load impedance be equal to the conjugate of the source impedance given by:

$$Z_{IN} = Z_S^*$$

For maximum power transfer, two cases are considered.

Case 1. When Source and Load Are Real Impedances

Figure 5-2 shows the case when source and load impedances are real.

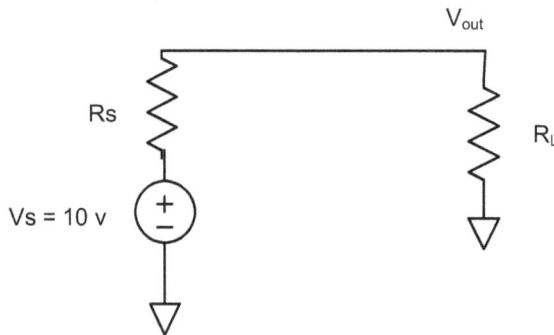

Figure 5-2 Network with purely resistive source and load impedance

Example 5-1: Use the Genesys linear simulation techniques to prove the maximum power transfer condition in a purely resistive network when $Z_S = Z_L = 50$ Ohm.

Solution: This is a very simple Genesys schematic with only input and output ports, as shown in Figure 5-3. The input port represents an RF signal source with 50 Ω source impedance and the output port is simply a 50 Ω resistive load. Use a Linear Analysis to sweep the schematic from 450 to 550 MHz and add a rectangular graph with the insertion loss, S21, in dB. Figure 5-3 shows the schematic and plot indicating that S21 = 0 dB. Because there is zero insertion loss between the source and load, there is maximum power transfer.

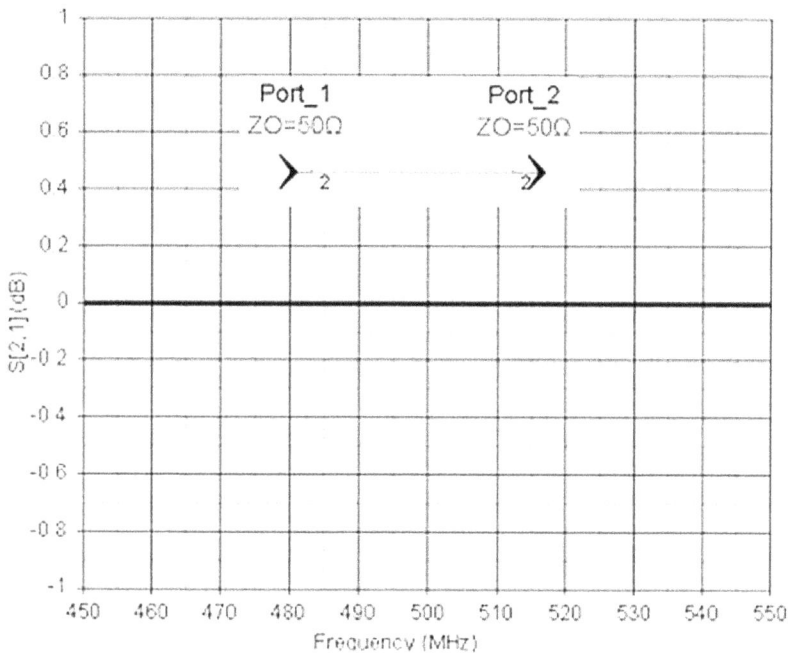

Figure 5-3 Case I power transfer with $R_L = R_S$

Notice that when the load impedance is equal to the source impedance the insertion loss is 0 dB indicating maximum power transfer.

Example 5-2: Use the Genesys linear simulation techniques to prove the maximum power transfer condition if the load is 25 Ohm.

Change the load impedance in Figure 5-3 to 25 Ω as shown in Figure 5-4. Simulate the schematic and display the insertion loss, S21, in a rectangular plot.

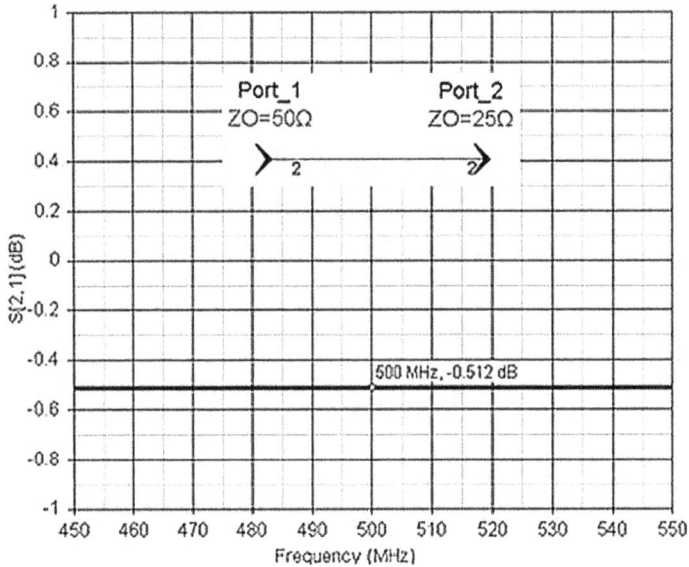

Figure 5-4 Power transfer with $R_L = 25$ Ohm

Notice that S21 = -0.512 dB indicating insertion loss. To calculate the amount of power loss, we know that the maximum power transfer is 0.5 Watts. To determine the amount of 0.512 dB power loss we first convert the 0.5 Watts to dBm which is the power in dB relative to 1 mW. Utilizing the conversion equation,

$$10 \log (\text{power in mW}) = \text{power in dBm}.$$

$$10 \log (500 mW) = 26.98 \ dBm$$

The loss of 0.512 dB is subtracted from the 26.98 dBm to result in (26.98 dBm − 0.512 dB) = 26.47 dBm. Then converting from dBm back to mW we get 444 mW as calculated in Example 5.2-1.

$$10^{\left(\frac{26.47}{10} \right)} = 444 \ mW \quad or \quad 0.444 \ Watts$$

This proves that the maximum power is transferred when $R_S = R_L$.

Case 2. When Load is a Complex Impedance

We know that maximum power transfer occurs when $Z_S = Z_L^*$. Therefore, if $Z_L = R_L - jX_L$, then for maximum power transfer we must have $Z_S = R_L + jX_L$.

Example 5-3: If the load is 50 Ω in series with a 15 pF capacitance, find the source impedance to have maximum power transfer at 500 MHz.

Solution: We can find a source inductance that cancels the reactance of the load at this frequency. The reactance of the 15 pF capacitor is:

$$X_C = \frac{1}{2\pi f C} = 21.231 \ \Omega$$

Therefore,

$$Z_L = 50 - j21.231 \ \Omega$$

For maximum power transfer the source impedance must be:

$$X_S = 50 + j21.231 \ \Omega$$

At 500 MHz the value of the source series inductor is:

$$L = \frac{21.231}{2\pi f} = 6.76 \ nH$$

To demonstrate the maximum power transfer, create a schematic in Genesys as shown in Figure 5-5. Place the specified 15 pF capacitance in series with the load and the 6.76 nH inductance in series with the source. Plot the insertion loss and VSWR on a rectangular graph as shown in Figure 5-5. Note that unlike the purely resistive source and load case, a complex conjugate match occurs at a single frequency. A perfect 1:1 VSWR is achieved at the conjugate match frequency of 500 MHz.

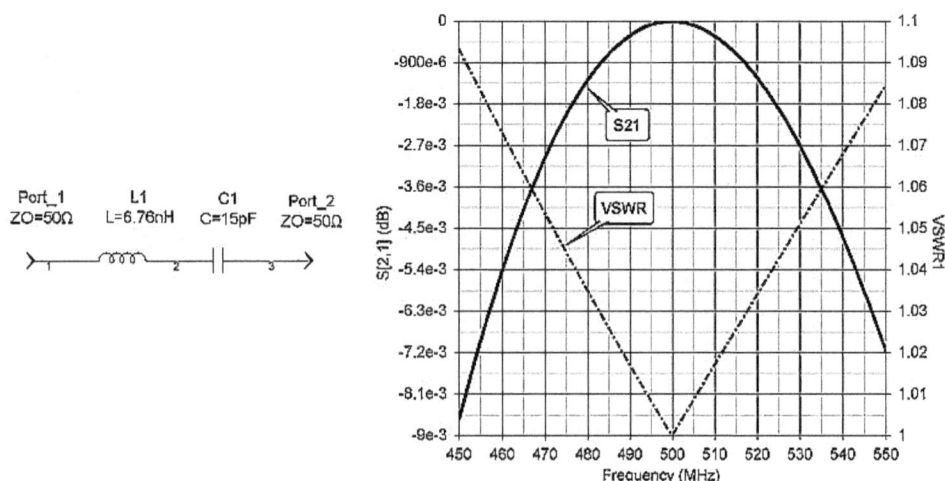

Figure 5-5 Maximum power transfer with complex source and load impedance

5.3 Matching a Complex Load to a Complex Source

One of the important tasks in RF and microwave engineering is the determination of how an arbitrary complex load impedance, $Z_L = R_L + jX_L$, is analytically matched to any complex source impedance, $Z_S = R_S + jX_S$, as shown in Figure 5-6. This problem arises mainly in the design of inter-stage matching networks between active devices or between an antenna and a transmitter.

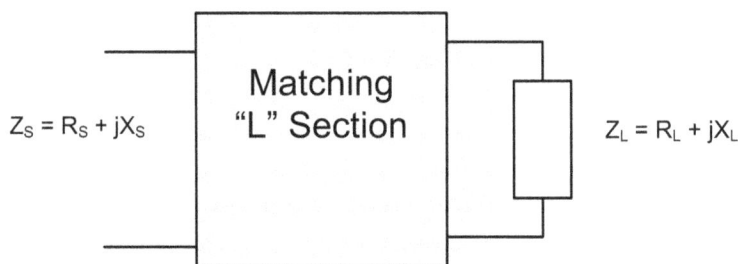

Figure 5-6 General impedance matching with an L-network

Example 5-4: Design L-networks matching a complex load impedance, $Z_L = 10 - j15 \, \Omega$, to a complex source impedance, $Z_S = 15 - j20 \, \Omega$, at a frequency of 2 GHz.

This problem has four solutions. The solutions are given separately.

First Solution: Use Equations (5-7) and (5-8) in the textbook to calculate the matching element values.

```
RS=15
XS=-20
RL=10
XL=-15
f=2e9
B1=((RL*XS)+sqrt(RS*RL*(RS^2+XS^2-RS*RL)))/(RL*(RS^2+XS^2))
X1=(RL*XS-RS*XL)/RS+(RS-RL)/(B1*RS)
L1=X1/(2*pi*f)
C2=B1/(2*pi*f)
```

Figure 5-7 Calculation of the matching element values

Viewing Workspace Variables show that $B1$= 0.011 and $X1$ = 32.795, therefore, the series element is an inductor, L1 = 2.61 nH, and the shunt element is a capacitor, $C2$ = 0.852 pF. The schematic and swept response is shown in Figure 5-8. The input return loss, S11, shows that the network is perfectly matched at 2 GHz and the measured fractional bandwidth at 20 dB return loss is about 11 %, indicating a narrow matching bandwidth. Fractional bandwidths less than 20% are generally considered narrow band. The schematic and response of the first solution is shown in Figure 5-8.

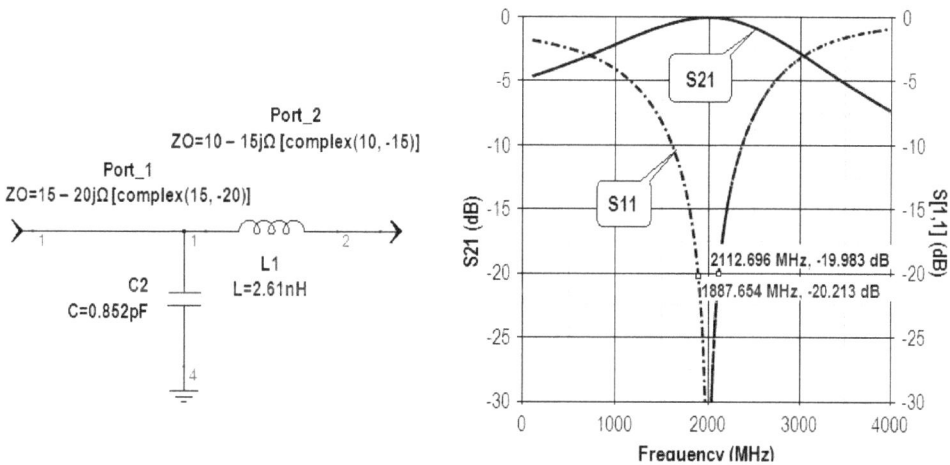

Figure 5-8 Schematic and response of the matching L-network

Example 5-5: Design the second L-network to match the complex load impedance, $Z_L = 10 - j15$ Ω, to a complex source impedance, $Z_S = 15 - j20$ Ω, at 2 GHz

Second Solution: Use Equations (5-9) and (5-10) to calculate the matching element values.

```
RS=15
XS=-20
RL=10
XL=-15
f=2e9

B2=((RL*XS)-sqrt(RS*RL*(RS^2+XS^2-RS*RL)))/(RL*(RS^2+XS^2))
X2=(RL*XS-RS*XL)/RS+(RS-RL)/(B2*RS)
C1=-1/(2*pi*f*X2)
L2=-1/(2*pi*f*B2)
```

Figure 5-9 Calculation of the matching element values

Viewing Workspace Variables show that *B2= -0.075* and *X2* = -2.795, therefore, the series element is a capacitor, *C1* = 28.47 pF and the shunt element is an inductor, *L2* = 1.065 nH. To plot the response of the matching network, create a new Design with Schematic and use the given element values. Simulate the schematic and display the return loss, S11, and insertion loss, S21, in dB. Notice that the input return loss, S11, in Figure 5-12 shows that the network is matched at 2 GHz and the fractional bandwidth at 20 dB return loss is about 12 % of the center frequency. The schematic and the swept response are shown in Figure 5-10.

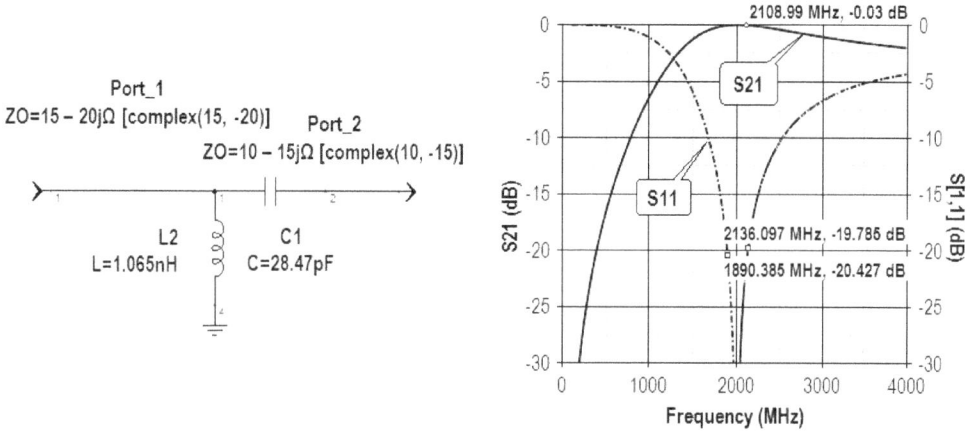

Figure 5-10 Schematic and response of the matching L-network

Example 5-6: Design the third L-network to match the complex load impedance, $Z_L = 10 - j15$ Ω, to a complex source impedance, $Z_S = 15 - j20$ Ω, at 2 GHz.

Third Solution: Use the textbook Equations (5-12) and (5-13) to calculate the matching element values.

```
RS=15
XS=-20
RL=10
XL=-15
f=2e9
B3=((RS*XL)+sqrt(RS*RL*(RL^2+XL^2-RS*RL)))/(RS*(RL^2+XL^2))
X3=(RS*XL-RL*XS)/RL+(RL-RS)/(B3*RL)
L1=-1/(2*pi*f*B3)
L2=X3/(2*pi*f)
```

Figure 5-11 Calculating the matching element values

Viewing the Workspace Variables show that *B3*= -0.013 and *X3* = 36.202, therefore, the shunt element is an inductor, *L1* = 6.16 nH and the series element is also an inductor, *L2* = 2.881 nH. To plot the response of matching network set up a new Design with Schematic using the given element values. Simulate the schematic and display the input return loss,

S11, and insertion loss, S21, in dB. The matching network and the swept responses are shown in Figure 5-12. Notice that the fractional bandwidth at 20 dB return loss is about 16 % of the center frequency indicating a narrowband matching network. The schematic and the swept response are shown in Figure 5-12.

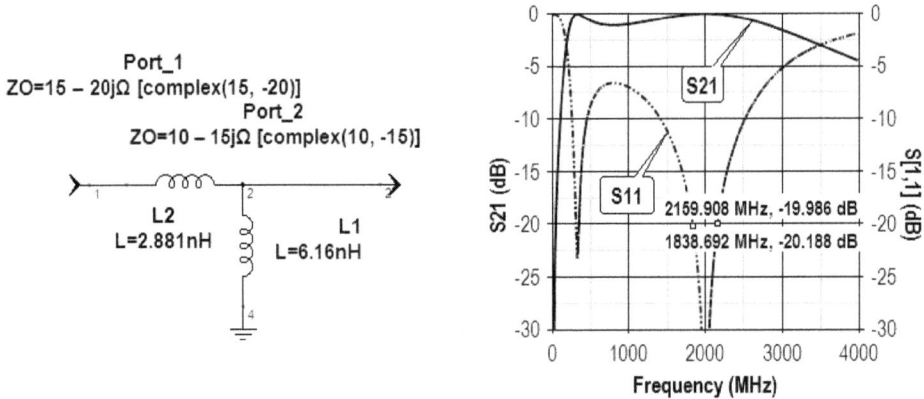

Figure 5-12 Solution and response of the matching network for the third solution

Example 5-7: Design the fourth L-network to match the complex load impedance, $Z_L = 10 - j15\ \Omega$, to a complex source impedance, $Z_S = 15 - j20\ \Omega$, at 2 GHz.

Fourth Solution: Use the textbook Equations (5-14) and (5-15) to calculate the matching element values.

```
RS=15
XS=-20
RL=10
XL=-15
f=2e9
B4=((RS*XL)-sqrt(RS*RL*(RL^2+XL^2-RS*RL)))/(RS*(RL^2+XL^2))
X4=(RS*XL-RL*XS)/RL+(RL-RS)/(B4*RL)
L1=-1/(2*pi*f*B4)
L2=X4/(2*pi*f)
```

Figure 5-13 Calculating the matching element values

Viewing Workspace Variables show that *B4*= -0.079 and *X4* = 3.798, therefore, the shunt element is a capacitor, *L1* = 1.002 nH and the series element is also an inductor, *L2* = 0.302 nH. To plot the response of the fourth solution, set up a new Design with Schematic in Genesys using the given element values. Simulate the schematic and display the Return Loss, S11, and insertion loss, S21, in dB. Notice that the input return loss, S11 in Figure 5-16 shows that the network is matched at 2 GHz and the measured fractional bandwidth at 20 dB return loss is about 15 % of the center frequency. The schematic and the swept response are shown in Figure 5-14.

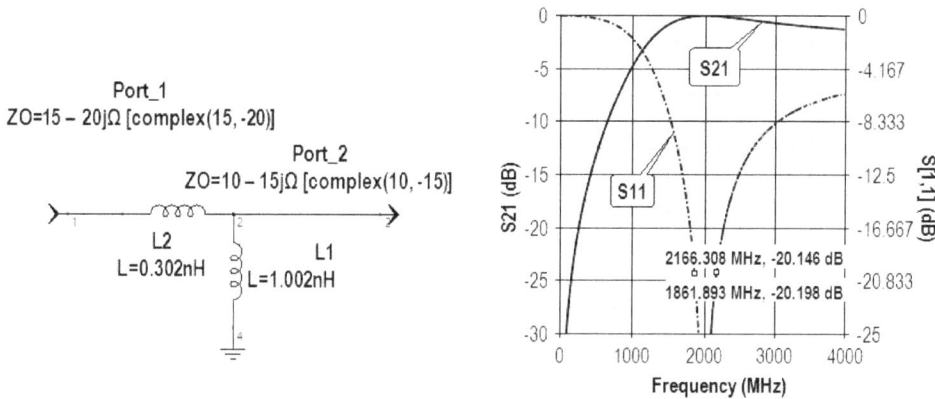

Figure 5-14 The matching network and response of the fourth solution

5.4 Matching a Complex Load to a Real Source

The Combined conditions for valid solutions are summarized in Table 5-1.

Case #	First Condition	Second Condition	# of Solutions	Equations Used
1	$R_L < R_S$	$R_L^2 + X_L^2 - R_L R_S > 0$	4	(5-21) to (5-24) (5-26) to (5-29)
2	$R_L < R_S$	$R_L^2 + X_L^2 - R_L R_S < 0$	2	(5-21) to (5-24)
3	$R_L > R_S$	N/A	2	(5-26) to (5-29)

Table 5-1 Impedance matching conditions and the number of solutions

The simplified version of these equations are given in the textbook.

Example 5-8: Design a single L-network that will match a real source impedance $Z_0 = 50\ \Omega$ to a complex load impedance, $Z_L = 7 - j22\ \Omega$, at a frequency of 1 GHz.

Notice that for this example $R_L < Z_0$ and $R_L^2 + X_L^2 - R_L Z_0 = 183 > 0$, therefore, the matching network has four solutions.

First Solution: Use Equations (5-32) and (5-33) to calculate the matching element values.

```
Z0=50
RL=7
XL=-22
f=1000e6
r=RL/Z0
x=XL/Z0
B1=sqrt((1-r)/r)/Z0
X1=Z0*sqrt(r*(1-r))-x*Z0
LS1=X1/(2*pi*f)
CP2=B1/(2*pi*f)
```

Figure 5-15 Calculation for matching element values

Viewing Workspace Variables show that the series element is an inductor, _LS1_ = 6.263 nH, and the shunt element is a capacitor, _CP2_ = 7.889 pF. To display the frequency response, setup a new Design with Schematic in Genesys and use the given element values. Simulate the schematic and display the input return loss, S11, and insertion loss, S21, in dB.

The schematic and response of the first solution is shown in Figure 5-16. The simulated response in Figure 5-16 shows that the load is perfectly matched to the source at 1 GHz and the measured fractional bandwidth at 20 dB return loss is about 5 % indicating a narrowband matching network.

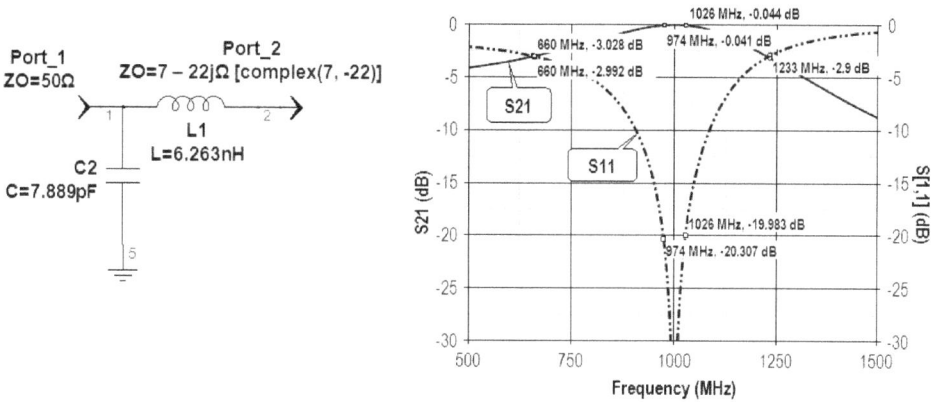

Figure 5-16 Schematic and response of the first solution

Example 5-9: Design the second L-network that will match a real source impedance $Z_0 = 50\ \Omega$ to a complex load impedance, $Z_L = 7 - \text{j}22\ \Omega$, at a frequency of 1 GHz.

Second Solution: Use Equations (5-34) and (5-35) in an Equation Editor to calculate the matching element values.

```
Z0=50
RL=7
XL=-22
f=1000e6
r=RL/Z0
x=XL/Z0
B2=-sqrt((1-r)/r)/Z0
X2=-Z0*sqrt(r*(1-r))-x*Z0
L1=X2/(2*pi*f)
L2=-1/(2*pi*f*B2)
```

Figure 5-17 Calculating the matching element values

Viewing Workspace Variables show that the series element is an inductor, $L1 = 0.740$ nH and the shunt element is another inductor $L2 = 3.211$ nH. To display the frequency response, set up a new Design with Schematic in Genesys and use the given element values. Simulate the schematic from 500 to 1500 MHz and display S11, and S21, in dB.

The schematic and simulated response in Figure 5-18 shows that the impedance matching network perfectly matches the complex load to a 50 Ω source resistor at 1 GHz. The measured fractional bandwidth at 20 dB return loss is about 13 % indicating a narrowband matching network.

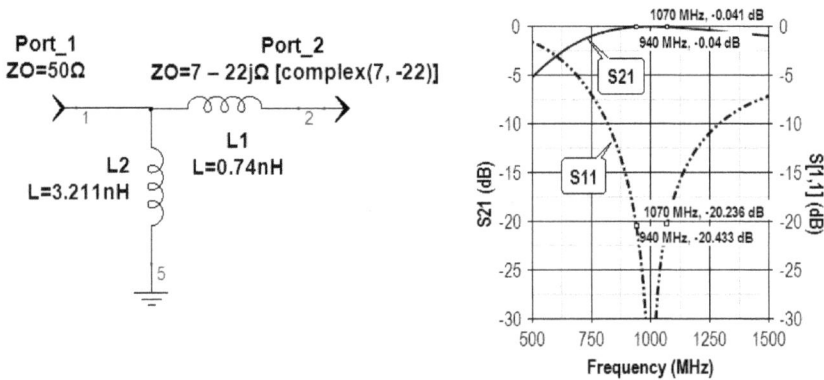

Figure 5-18 Schematic and response of the second solution

Example 5-10: Design the third L-network that will match a real source impedance $Z_0 = 50$ Ω to a complex load impedance, $Z_L = 7 - j22$ Ω, at a frequency of 1 GHz.

Third Solution: Use Equations (5-36) and (5-37) in an Equation Editor to calculate the matching element values.

```
Z0=50
RL=7
XL=-22
f=1000e6
r=RL/Z0
x=XL/Z0
B3=(x+sqrt(r*(r^2+x^2-r)))/(Z0*(r^2+x^2))
X3=Z0*sqrt((r^2+x^2-r)/r)
L2=X3/(2*pi*f)
L1=-1/(2*pi*f*B3)
```

Figure 5-19 Calculating the matching element values

Viewing Workspace Variables show that the shunt element is an inductor, $L1 = 5.008$ *nH* and the series element is another inductor, $L2 = 5.754$ *nH*.

To display the frequency response, setup a new Design with Schematic in Genesys and use the given element values. Simulate the schematic and display the input return loss and insertion loss in dB.

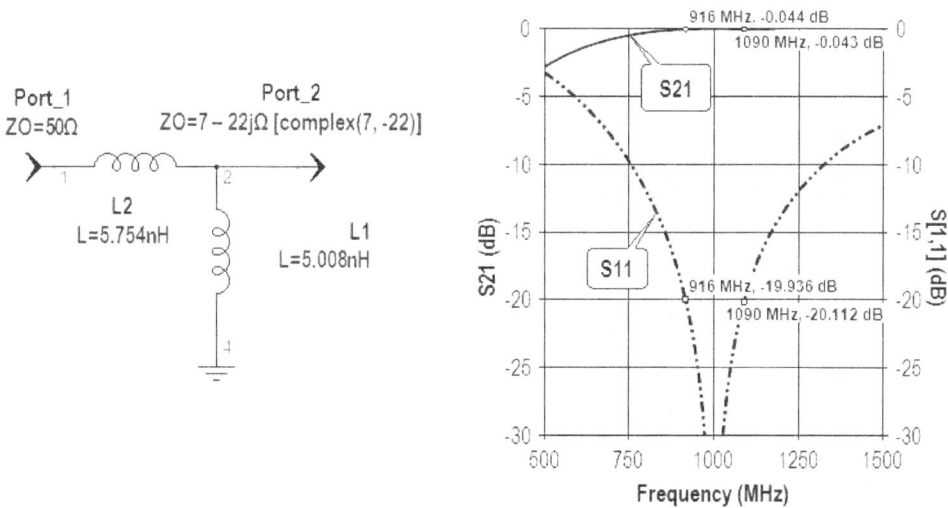

Figure 5-20 Schematic and response of the third solution

The simulated response in Figure 5-20 shows that the load is perfectly matched to the 50 Ω source resistor at 1 GHz and the measured fractional bandwidth at 20 dB return loss is about 17 %.

Example 5-11: Design the fourth L-network that will match a real source impedance $Z_0 = 50\ \Omega$ to a complex load impedance, $Z_L = 7 - j22\ \Omega$, at a frequency of 1 GHz.

Fourth Solution: Use Equations (5-38) and (5-39) in an Equation Editor to calculate the matching element values.

```
Z0=50
RL=7
XL=-22
f=1000e6
r=RL/Z0
x=XL/Z0
B4=(x-sqrt(r*(r^2+x^2-r)))/(Z0*(r^2+x^2))
X4=-Z0*sqrt((r^2+x^2-r)/r)
C2=-1/(2*pi*f*X4)
L1=-1/(2*pi*f*B4)
```

Figure 5-21 Calculating the matching element values

Viewing the Workspace Variables show that the series element is an inductor, *L1 = 3.135 nH* and the shunt element is a capacitor, *C2 = 4.402 pF*. To display the frequency response, setup a new Design with Schematic in Genesys and use the given element values. Simulate the schematic and display S11, and S21, in dB.

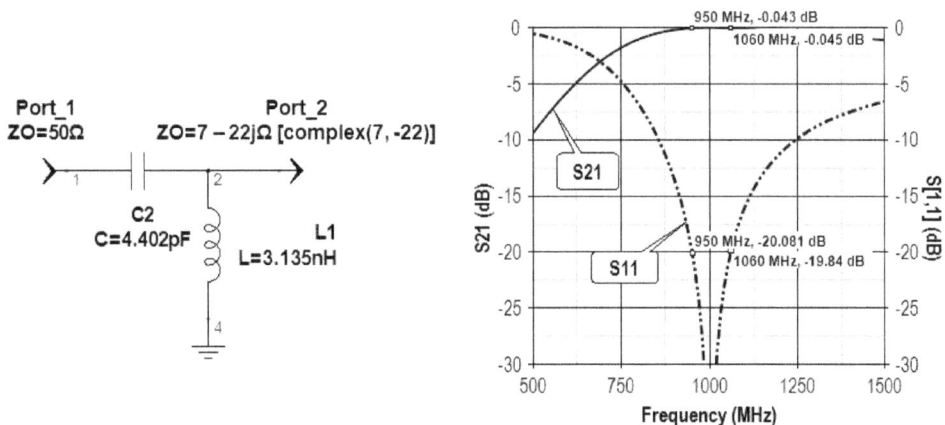

Figure 5-22 Schematic and response of the fourth solution

The simulated response in Figure 5-22 shows that the matching network matches the complex load impedance to a 50 Ω source resistor at 1 GHz.

5.5 Matching a Real Load to a Real Source Impedance

When source and load impedances are both real the first matching configuration is redrawn in Figure 5-23.

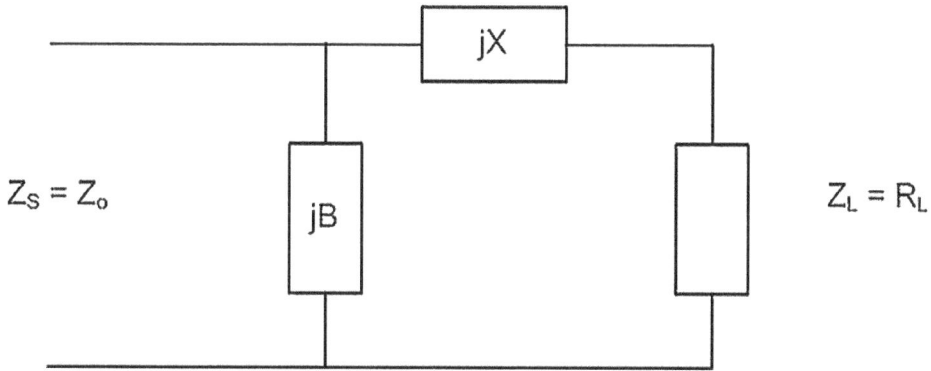

Figure 5-23 First matching configuration with $X_L = X_S = 0$

The two solutions are given by Equations (5-47) through (5-50).

Case No	Condition	Solutions	Use Equations
1	$r < 1$	2	(5-42) to (5-45)
2	$r > 1$	2	(5-47) to (5-50)

Table 5-2 Impedance matching conditions and the number of solutions

Example 5-12: Design the first L-network to match a 10 Ω load to a 50 Ω source resistor at 500 MHz.

First Solution: Since the load resistor is smaller than the source resistor, we get two solutions from Equations (5-42) through 5-45 in the Equation Editor.

```
Z0=50
RL=10
f=500e6
r=RL/Z0
B1=sqrt((1-r)/r)/Z0
X1=Z0*sqrt(r*(1-r))
L1=X1/(2*pi*f)
C2=B1/(2*pi*f)
```

Figure 5-24 Calculation of the matching element values

To design the first matching L-network, create a new Design with Schematic in Genesys and select the element values by viewing the Workspace Variables. Simulate the schematic from 0 to 1000 MHz and display the response, as shown in Figure 5-25

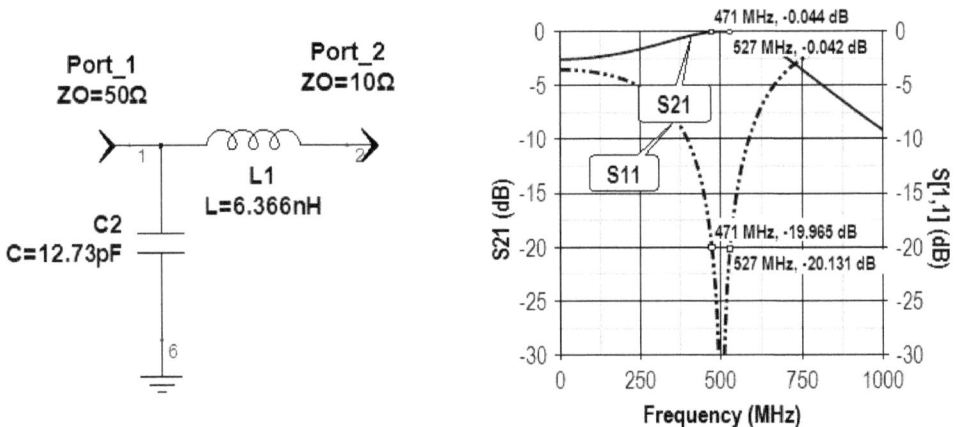

Figure 5-25 Schematic and response of the first matching L-network

The simulated response in Figure 5-25 shows that the matching L-network perfectly matches the 10 Ω load resistor to a 50 Ω source resistor at 500 MHz and the fractional bandwidth at 20 dB return loss is about 11 %.

Example 5-13: Design the second L-network to match a 10 Ω load to a 50 Ω source resistor at 500 MHz.

Second Solution: To design the matching L-network, use Equations (5-44) and (5-45) in the Equation Editor to calculate the matching element values.

```
Z0=50
RL=10
f=500e6
r=RL/Z0
B2=-sqrt((1-r)/r)/Z0
X2=-Z0*sqrt(r*(1-r))
C1=-1/(2*pi*f*X2)
L2=-1/(2*pi*f*B2)
```

Figure 5-26 Calculating the matching element values

To display the frequency response, setup a Design with Schematic in Genesys and select the element values by viewing the Workspace Variables. Simulate the schematic from 0 to 1000 MHz and display the response. The schematic and response of the second solution is shown in Figure 5-27. Note that the load resistor is perfectly matched to the source at 500 MHz. The measured fractional bandwidth at 20 dB return loss is about 11 %.

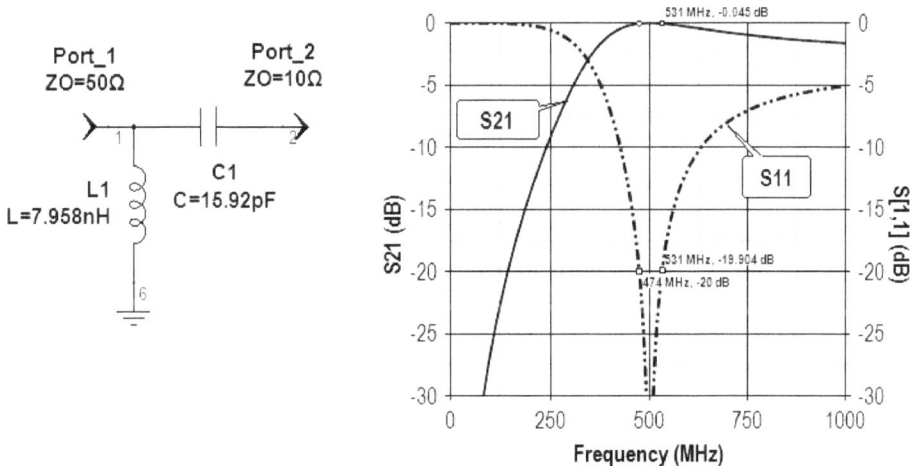

Figure 5-27 Schematic and response of the second matching L-network

The simulated response in Figure 5-27 shows that the matching L-network perfectly matches the load impedance to the source impedance at 0.5 GHz.

5.6 Introduction to Broadband Matching Networks

In the previous sections, the L-section matching networks achieved an impedance match at a fixed frequency capable of producing a 20 dB return loss over a narrow fractional bandwidth of less than 20%. Broadband networks are generally considered to have greater than 20% fractional bandwidths. In this section it is demonstrated that the bandwidth of a matching network can be increased by cascading L-networks. It is demonstrated that by cascading L-networks of equal Q factor, the bandwidth of a network can be increased. The design of equal-Q matching networks is based on the selection of intermediate, or virtual, resistors not necessarily 50 Ω, and then matching the load and source impedance to the virtual resistors. Successively adding additional L-networks of equal Q will continue to extend the bandwidth of the overall circuit.

5.7 Broadband Impedance Matching Design

This section demonstrates the importance of the selection of the proper intermediate network resistance that will result in the best broadband return loss. In example 5.4-1 the complex source and load impedance are matched to one specific intermediate resistor thus creating two, equal-Q, L-networks. The purpose is to show that this method provides a broader matching bandwidth at 20 dB return loss compared to the case when we chose a different intermediate resistor.

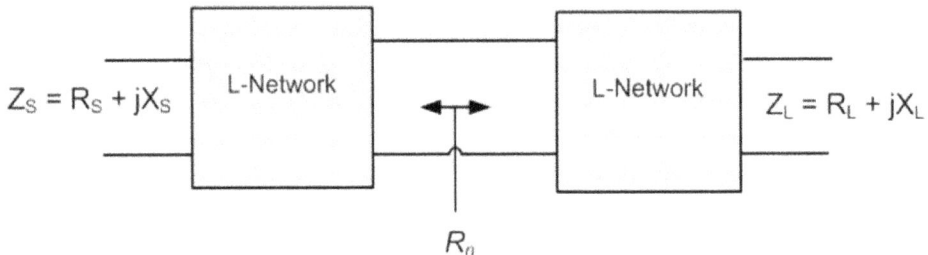

Figure 5-28 Cascaded L-networks with intermediate resistance, R_n

Example 5-14: A transmitter operates over a frequency range of 835 MHz to 1200 MHz. At its center frequency of 1 GHz the source impedance is Z_S

= $55 + j10 \ \Omega$ while the antenna input impedance is ZL = $20 + j15 \ \Omega$. Design a cascade of two L-networks that matches the transmitter output to the antenna impedance.

Solution: The first step in the solution is to calculate the intermediate resistor, Rn. The optimum intermediate resistor value is equal to square root of the product of the real part of the source and load impedance.

$$R_n = \sqrt{(55)(20)} = 33.166 \ \Omega$$

Use Equations (5-32) and (5-33) in an Equation Editor to calculate the element values of the load matching network. Viewing the Workspace Variables in Figure 5-29 show that the series element is an inductor, LS1 = 0.195 nH and the shunt element is a capacitor, CP2 = 3.893 pF.

```
Z0=33.166
RL=20
XL=15
f=1000e6
r=RL/Z0
x=XL/Z0
B1=sqrt((1-r)/r)/Z0
X1=Z0*sqrt(r*(1-r))-x*Z0
LS1=X1/(2*pi*f)
CP2=B1/(2*pi*f)
```

Figure 5-29 Calculation of matching element values

To display the frequency response of the L-network, set up a new Design with Schematic in Genesys and select the element values as given in Figure 5-35. Simulate the schematic from 0 to 2000 MHz and display the input return loss, S11, and the insertion loss, S21, in dB, as shown in Figure 5-30.

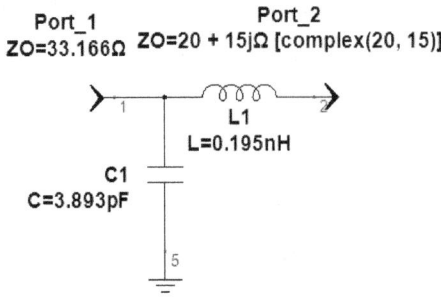

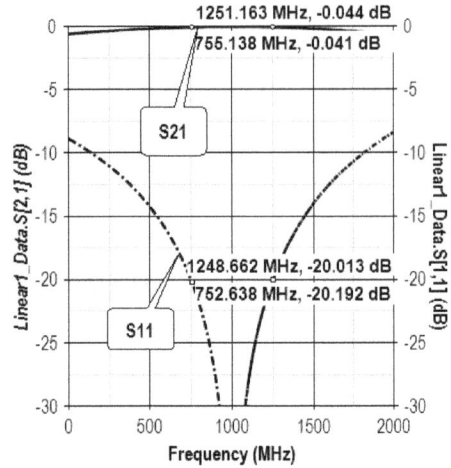

Figure 5-30 Schematic and response of the matching L-network

To match the source impedance to the intermediate resistor, we use Equations (5-36) and (5-37) in an Equation Editor and calculate the element values of the matching network. Viewing the Workspace Variables in Figure 5-31 show that the series element is an inductor, *LS1* = 4.458 nH and the shunt element is a capacitor, *CP2* = 2.875 pF.

```
Z0=33.166
RL=55
XL=10
f=1000e6
r=RL/Z0
x=XL/Z0
B3=(x+sqrt(r*(r^2+x^2-r)))/(Z0*(r^2+x^2))
X3=Z0*sqrt((r^2+x^2-r)/r)
LS1=X3/(2*pi*f)
CP2=B3/(2*pi*f)
```

Figure 5-31 Calculating the matching element values

To display the frequency response, set up a new Design with Schematic in Genesys and simulate the schematic from 0 to 2000 MHz. The input return loss, S11, and the insertion loss, S21, in dB are shown in Figure 5-32.

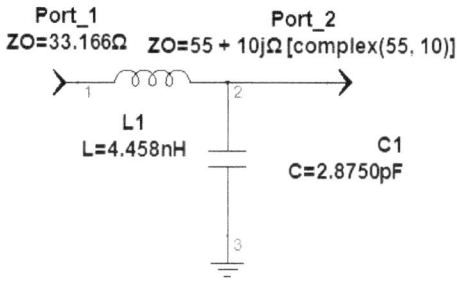

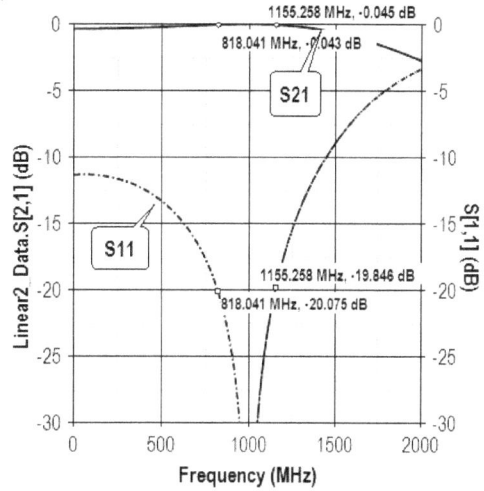

Figure 5-32 Schematic and response of the matching L-network

Finally, we cascade the two L-networks in Figure 5-33 and simulate the schematic to display the return loss and insertion loss in dB. To measure the matching bandwidth manually, place two markers at 20 dB return loss and record the marker readings. The schematic and simulated response of the cascaded network is shown in Figure 5-33.

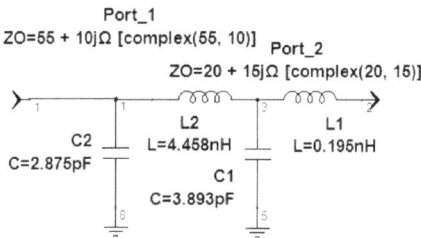

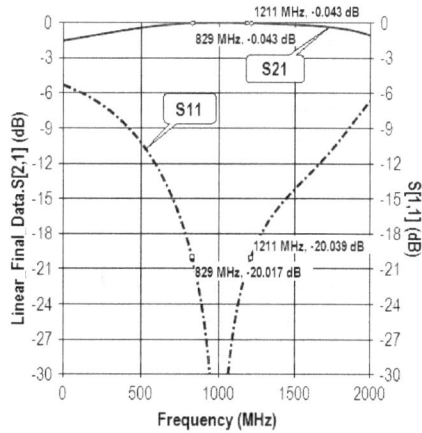

Figure 5-33 Cascaded L-networks schematic and response

The simulated response in Figure 5-42 shows that the cascaded network perfectly matches the complex load impedance to the source impedance at 1 GHz.

Notice that the matching bandwidth at 20 dB return loss is:

$$BW_{20\,dB\ return\ loss} = 1211 - 829 = 382 \quad MHz$$

Therefore, the corresponding fractional bandwidths at 20dB return loss is:

$$FBW_{20\,dB\ return\,loss} = \frac{382}{\sqrt{(1211)(829)}} = 0.381 = 38.1\ \%$$

Notice that the higher fractional bandwidth is due to cascading two equal Q L-networks. The Q factor of a single L-network is calculated by the following equation:

$$Q = \frac{1}{2}\sqrt{\frac{R_2}{R_1} - 1} \qquad\qquad R_2 > R_1$$

Where: R_2 and R_1 are the input and output resistors of the L-network.

For the Example 5.4-1 the Q factors of the two L-networks are equal:

$$Q_{source} = Q_{load} = \frac{1}{2}\sqrt{\frac{33.166}{20} - 1} = \frac{1}{2}\sqrt{\frac{55}{33.166} - 1} = 0.405$$

If the Q factors of the two cascading L-networks are not equal, the fractional bandwidth would be smaller. Example 5.4-2 proves this point.

Example 5-15: Redesign the matching network of Example 5.4-1 by matching the source and load impedances to an intermediate resistor of 50 Ω instead of 33.166 Ω. Compare the fractional bandwidth and Q factors of the two Examples.

Solution: To redesign the matching network, change the intermediate resistor in the Equation Editors to 50 Ω and recalculate the matching element values. The schematics shown in Figure 5-34. are the result of analysis performed in the Genesys workspace for Example 7-15.

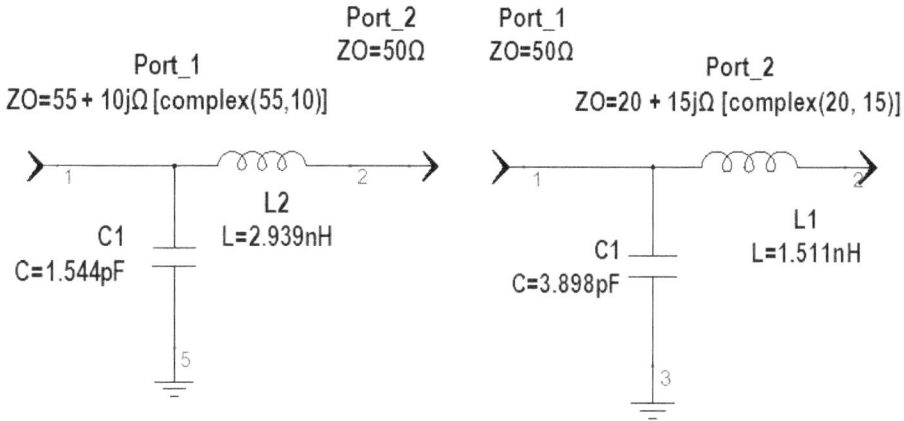

Figure 5-34 Load and source impedance matched to 50 Ω

Notice that the element values in both schematics have changed due to the change in the intermediate resistor. The cascaded schematic and simulated response are shown in Figure 5-35.

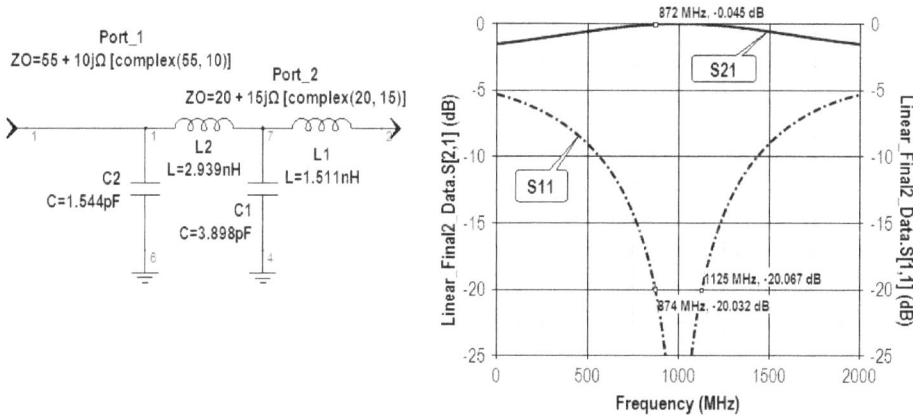

Figure 5-35 Complete matching network schematic and response

The markers at 20 dB return loss show that the bandwidth is:

$$BW_{20dB\ return\ loss} = 1125 - 874 = 251\ MHz$$

Therefore, the corresponding fractional bandwidths at 20dB return loss is:

$$FBW_{20dB\ return\ loss} = \frac{251}{\sqrt{(1125)(874)}} = 0.253 = 25.3\ \%$$

A comparison of the fractional bandwidths for the two examples show that the matching network in Figure 5-38 provides over 50 % more matching bandwidth than the matching network in Figure 5-40.

Now we show that the less bandwidth is due to the fact that the source and load matching networks do not have the same Q factors.

$$Q_{source} = \frac{1}{2}\sqrt{\frac{55}{50}-1} = 0.158$$

$$Q_{load} = \frac{1}{2}\sqrt{\frac{50}{20}-1} = 0.612$$

The equal Q factors in Example 5.4-1 cause the cascaded network to transfer more power to the load than the unequal Q factors in example 5.4-2.

5.8 Cascaded Broadband Impedance Matching

In Example 5-15 we showed that cascading two equal-Q matching L-networks increases the overall matching bandwidth. Therefore, we can further increase the matching bandwidth by increasing the number of equal-Q matching networks. In practice as the number of L-networks becomes greater than five, the element values become difficult to physically realize. In this section we show that by using a network of four equal-Q L-networks we can lower the individual Q factor and provide an even greater bandwidth than Example of 5-14.

Example 5-16: Redesign the matching network of Example 5-14 with four equal-Q L-networks. Compare the Q and bandwidth of Example 5-16 with Example 5-14.

Solution: This example uses the same steps developed in Example 5-14. The general equation for the calculation of any number of intermediate resistors is given by Equation (5-51).

$$R_n = R_S (r)^{n/N} \qquad n = 1, 2, 3, to \ N-1$$

Where:

R_n is the intermediate resistor values
R_S is the source resistor value
r is the normalized load resistor, $r = R_L/R_S$
N is the number of cascading networks

Equation Editor in Figure 5-36 is used to calculate the 3 intermediate resistors.

```
RS=55
RL=20
r=RL/RS
N=4
R1=RS*r^(1/N)
R2=RS*r^(2/N)
R3=RS*r^(3/N)
```

Figure 5-36 Calculation of the intermediate resistors: R1, R2, and R3

Viewing the Workspace Variables show that $R1 = 42.71\ \Omega$, $R2 = 33.166\ \Omega$, and $R3 = 25.755\ \Omega$. Now we design the four matching L-networks.

Equatin Editor in Figure 5-37 is used to calculatet the matching L-network between R3 and the load impedance .

```
Z0=25.755
RL=20
XL=15
f=1000e6
r=RL/Z0
x=XL/Z0
B1=sqrt((1-r)/r)/Z0
X1=Z0*sqrt(r*(1-r))-x*Z0
CS1=-1/(2*pi*f*X1)
CP2=B1/(2*pi*f)
```

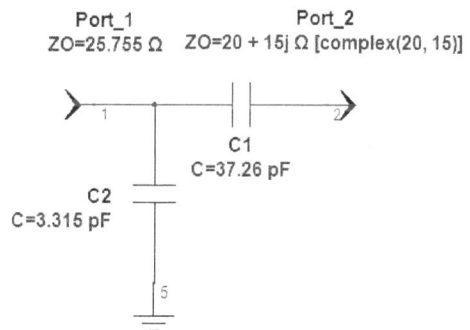

Port_1   Port_2
ZO=25.755 Ω   ZO=20 + 15j Ω [complex(20, 15)]

C1
C=37.26 pF

C2
C=3.315 pF

Figure 5-37 Calculation of the first matching L-network and schematic

Equatin Editor in Figure 5-38 is used to calculatet the matching L-network between R2 and R3.

```
Z02=33.166
RL2=25.755
f2=1000e6
r2=RL2/Z02
B12=sqrt((1-r2)/r2)/Z02
X12=Z02*sqrt(r2*(1-r2))
LS12=X12/(2*pi*f)
CP22=B12/(2*pi*f)
```

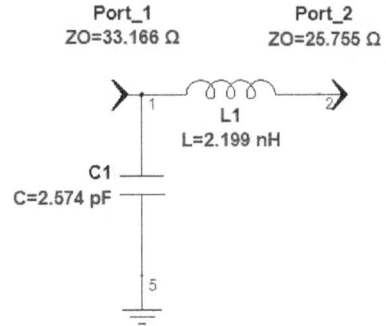

Figure 5-38 Calculation of the second matching L-network and schematic

Equatin Editor in Figure 5-39 is used to calculatet the matching L-network between R1 and R2.

```
Z03=42.71
RL3=33.166
f3=1000e6
r3=RL3/Z03
B13=sqrt((1-r3)/r3)/Z03
X13=Z03*sqrt(r3*(1-r3))
LS13=X13/(2*pi*f)
CP23=B13/(2*pi*f)
```

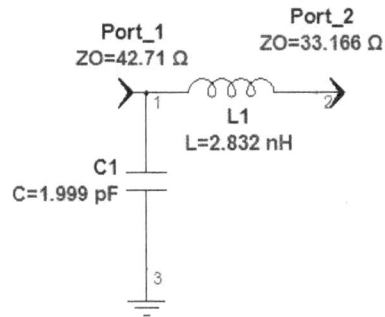

Figure 5-39 Calculation of the third matching L-network and schematic

Finally, Equatin Editor in Figure 5-40 is used to calculatet the matching L-network between R1 and the source impedance .

```
Z04=42.71
RL4=55
XL4=10
f4=1000e6
r4=RL4/Z04
x4=XL4/Z04
B34=(x4+sqrt(r4*(r4^2+x4^2-r4)))/(Z04*(r4^2+x4^2))
X34=Z04*sqrt((r4^2+x4^2-r4)/r4)
LS14=X34/(2*pi*f)
CP24=B34/(2*pi*f)
```

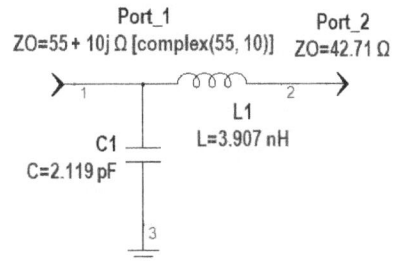

Figure 5-40 Calculation of the fourth matching L-network and schematic

Cascade the four L-networks and identify the nodes, from right to left, as shown in Figure 5-41.

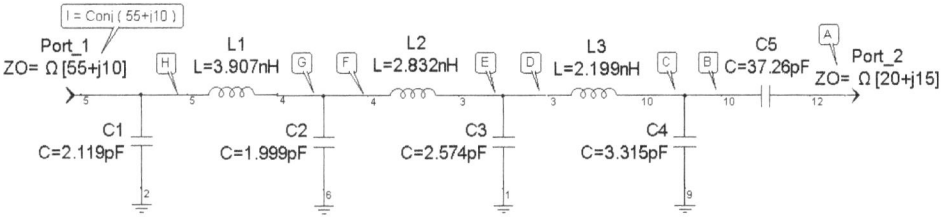

Figure 5-41 Schematic of the matching network using 4 L-networks

The simulated response shows that the matching bandwidth at 20 dB return loss is:

$$BW_{20dB\ return\ loss} = 1334 - 821 = 513\ MHz$$

Therefore, the fractional bandwidth at 20 dB return loss is:

$$FBW_{20dB\ return\ loss} = \frac{513}{\sqrt{(1334)(821)}} = 0.490 = 49.0\%$$

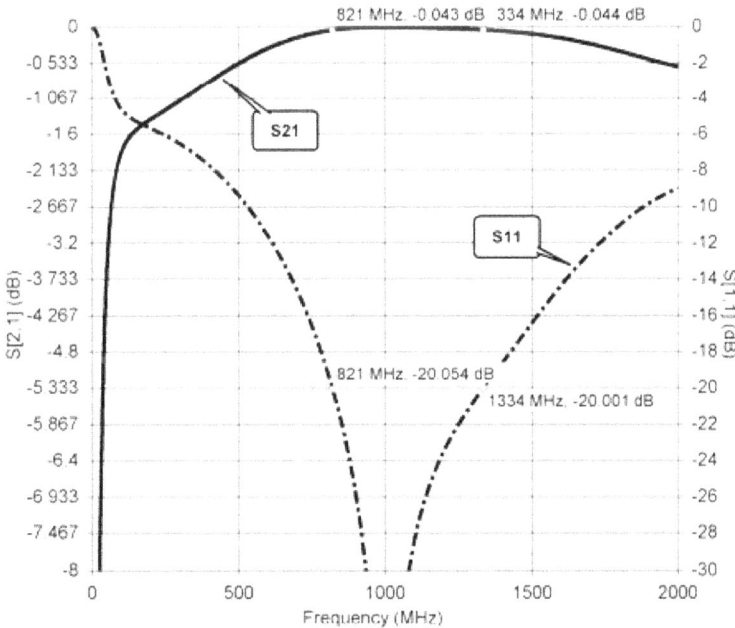

Figure 5-42 Response of the matching network using 4 L-networks

The comparison between the fractional bandwidth of this example with the fractional bandwidth of example 5.4-1 shows that cascading four equal-Q matching networks increases the fractional bandwidth by more than 93 % over the two equal-Q matching networks. As we stated earlier, the reason for wider matching bandwidth is that all four individual matching networks in Figure 5-46 have the same Q factor as shown in the following calculations.

$$Q_{4L} = \frac{1}{2}\sqrt{\frac{55}{42.71}-1} = \frac{1}{2}\sqrt{\frac{42.71}{33.166}-1} = \frac{1}{2}\sqrt{\frac{33.166}{25.755}-1} = \frac{1}{2}\sqrt{\frac{25.755}{20}-1} = 0.268$$

5.9 Limitations of Broadband Matching

From the previous example of section 5.5 we can deduce that the Q of the complex source and load impedance has an impact on the maximum bandwidth over which a good impedance match, low return loss, can be achieved. There is a finite limit on the achievable return loss for a given load impedance and circuit bandwidth known as Fano's limit [1]. Fano's Limit is the optimum reflection coefficient that can be achieved with a given load impedance. It is a theoretical limit that considers an infinite number of lossless matching elements. Fano's limit is defined by Equation (5-54).

$$\left| \Gamma \right| = e^{\frac{-\pi Q_L}{Q_{UL}}} \tag{5-54}$$

Where:

Q_{UL} = the unloaded Q or the ratio of the reactance to resistance of the load
Q_L = the loaded Q of the network or the ratio of the center frequency of the network divided by the 3 dB bandwidth as defined by Equation (4-11) in chapter 4 section 4.2.

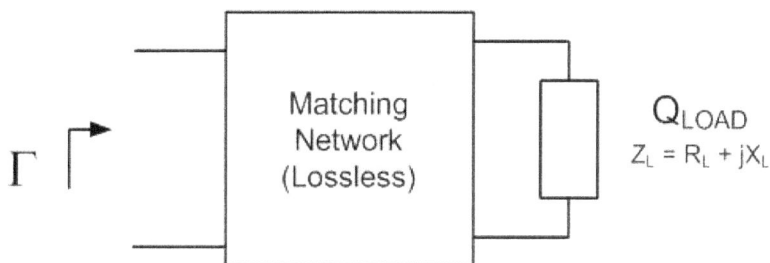

Figure 5-43 General configuration of impedance matching network

5.10 Example of Fano's Limit Calculation

In the following example we apply the Fano's limit to calculate the optimum reflection coefficient for a given network.

Example 5-17: Use the Genesys Impedance Matching Utility to match a source impedance of 100 Ω to a load that is comprised of a 500 Ω resistor in parallel with a 2 pF capacitance as shown in Figure 5-44. It is required that a 20 dB return loss and greater than -1 dB insertion loss be achieved over a frequency range of 60 MHz to 164 MHz.

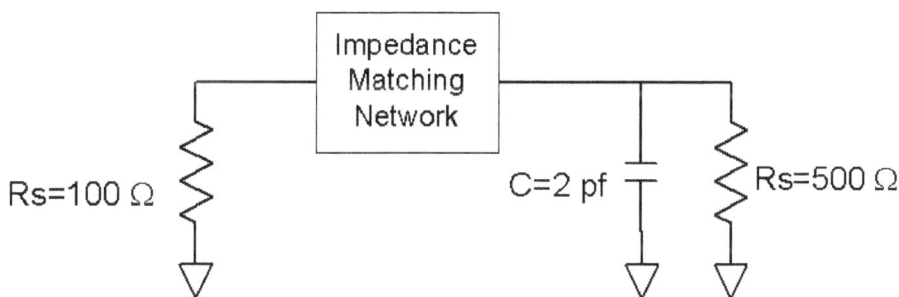

Figure 5-44 Match the 500 Ω and 2 pF loads to the 100 Ω source impedance

Solution: The impedance matching parameters are defined in the Match Properties window. On the Setting Tab, the lower and upper passband frequencies can be entered. Also enter the desired number of points to be used in the simulation. On the Sections Tab, the source and load impedances are defined by the selection of the network type as shown in Figure 5-60. Also on the Sections Tab, the type of network type and order can be selected. Figure 5-45 shows the variety of network types that can be selected. Basic LC tee networks can be used to model the L-networks as designed in section 5.3. Because the goal of this design is to have a bandpass characteristic the LC Bandpass network type is selected. To keep the network realized of all LC elements, select 'no transformers' to be used in the matching network. Finally select the order of the network and press Calculate Match button. A schematic of the synthesized network along with its response will be automatically generated. It is very easy to perform 'what-if' analysis using the Matching Utility. Simply enter a higher network order and select 'Calculate' to quickly compare the return loss.

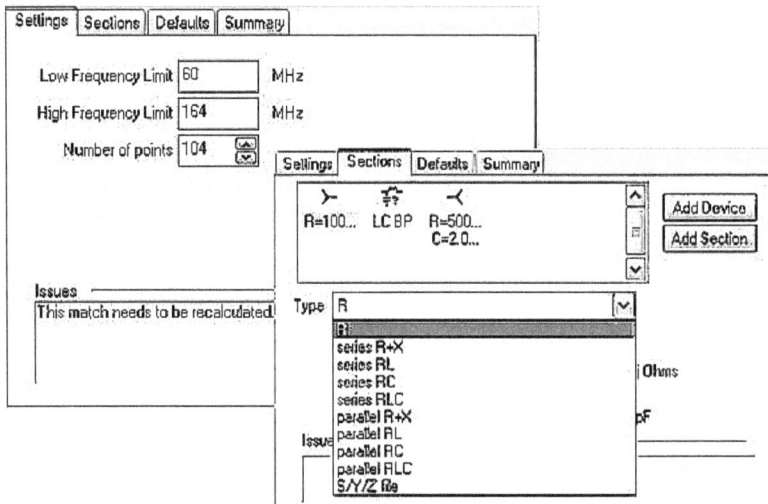

Figure 5-45 Frequency band, input and output impedance

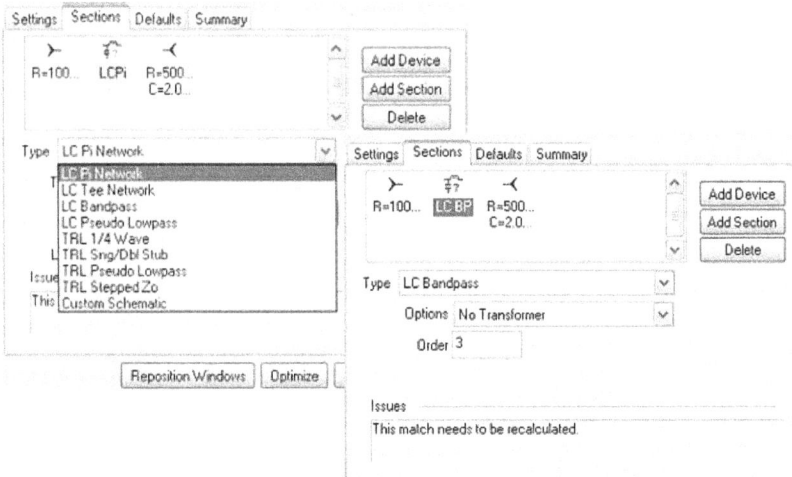

Figure 5-46 Specification of the network type and order

Figure 5-47 shows the return loss comparison between a third order and fifth order matching network. To meet the 20 dB return loss specification the fifth order network should be chosen.

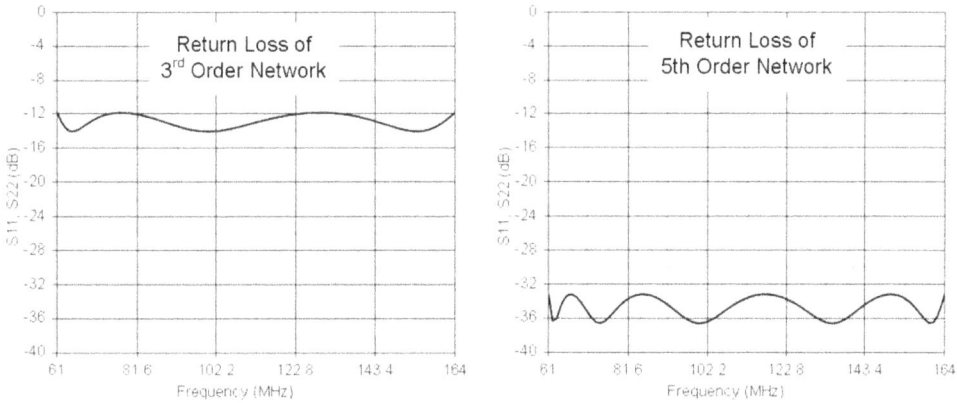

Figure 5-47 Return loss of third and fifth order matching networks

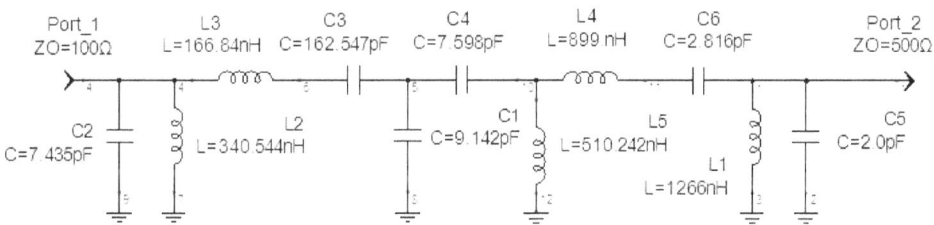

Figure 5-48 schematic of the fifth order impedance matching network

5.11 Effect of Finite Q on the Matching Networks

To complete the matching circuit design, the components need to be converted to their physical equivalents by accounting for the finite Q factor of the capacitors and inductors. This can be accomplished by using either S parameter or Modelithic models for the elements. A quick approach is to model each element with a Q factor that closely approximates the physical components. Note that a Q factor can be specified for each element in the Matching synthesis utility. In this design we'll demonstrate a technique to assign a common variable name representing the Q factor of each component. This allows the evaluation of the necessary Q factor of the components required to meet the specifications of the matching network. The variable name, Qcap, is assigned to all capacitors in the matching network. Similarly the variable name, Qind, is assigned to all of the inductors. The frequency at which the Q factor is defined is also defined as a variable, Qf. This allows the Q factor of all inductors and all capacitors to be tuned simultaneously.

Assign Qcap and Qind values to capacitors and inductors in Figure 5-49. It is required that a 20 dB return loss and greater than -1 dB insertion loss be achieved over a frequency range of 60 MHz to 160 MHz

The network response with the initial Qcap = 150 and Qind = 100 values is shown in Figure 5-50. We can see that the insertion loss (>-1dB) and return loss (<20 dB) specifications have been met. Figure 5-49 shows the revised schematic with the Q factor defined for all components. The variables are then initialized in the Equation Editor. Note the use of the (=?) which enables the variable to be tuned.

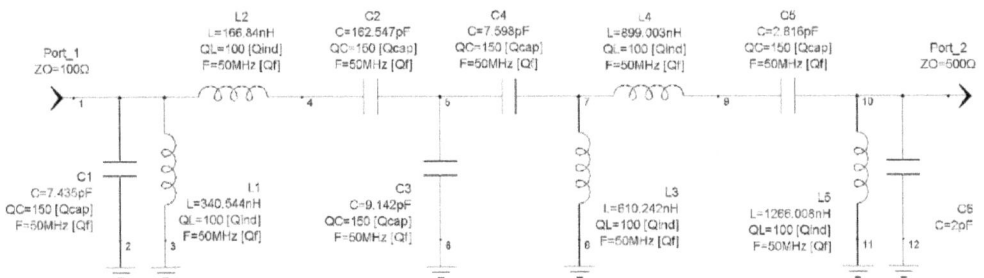

Figure 5-49 Fifth order matching schematic with Q factor variables

The response with Qcap=50 and Qind=40 is shown in Figure 5-51

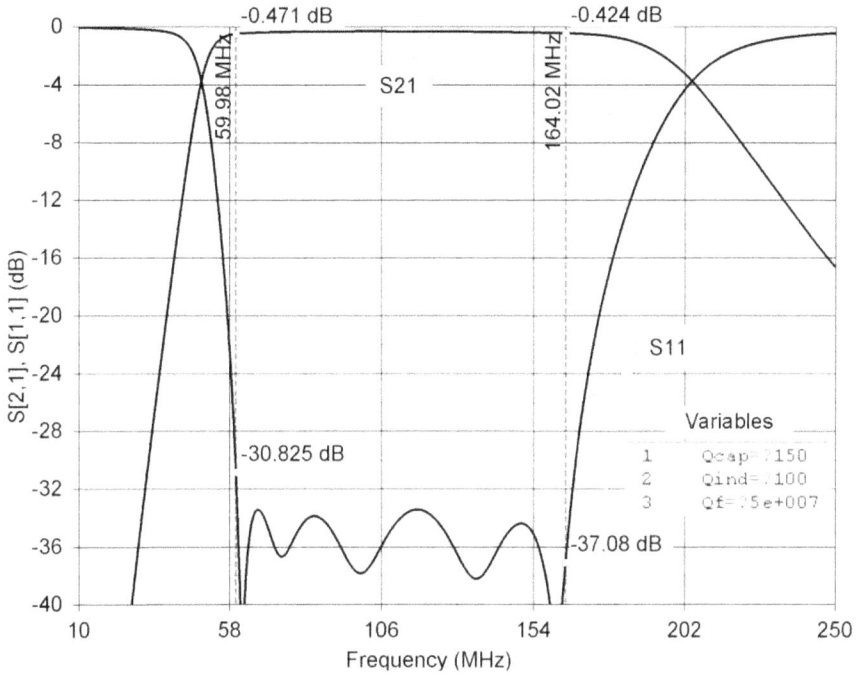

Figure 5-50 Response with Qcap=150 and Qind=100

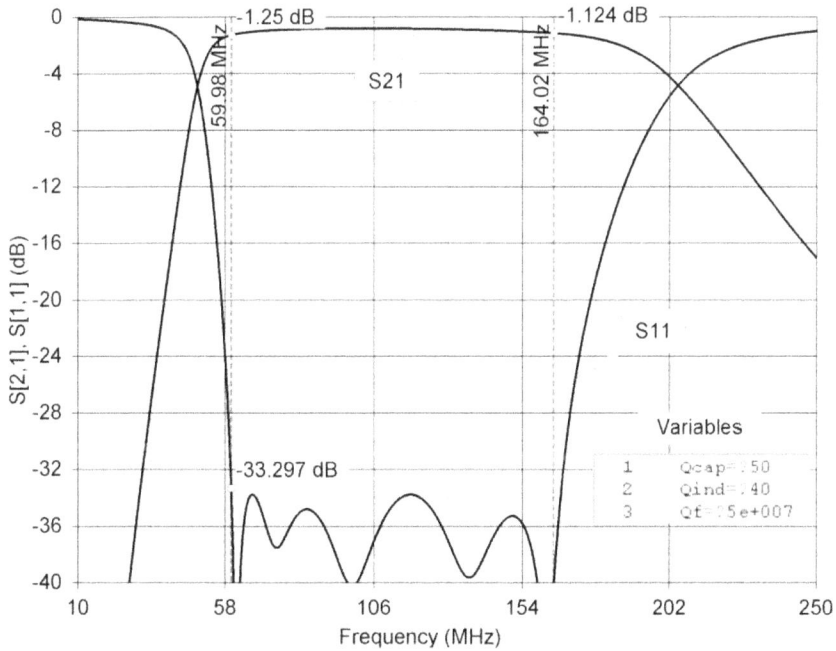

Figure 5-51 Response with Qcap=50 and Qind=40

References and Further Reading

[1]   Ali A. Behagi and Stephen D. Turner, *Microwave and RF Engineering,* A Simulation Approach with Keysight Genesys Software, BT Microwave LLC, March, 2015

[2]   Keysight Technologies, *Genesys 2015.08, Users Guide,* www.keysight.com

[3]   Guillermo Gonzales, *Microwave Transistor Amplifiers – Analysis and Design,* Second Edition, Prentice Hall Inc., Upper Saddle River, NJ.

[4]   Randy Rhea, *The Yin-Yang of Matching: Part 2 – Practical Matching Techniques,* High Frequency Electronics, March 2006

[5]   Steve C. Cripps, *RF Power Amplifiers for Wireless Communications,* Artech House Publishers, Norwood, MA. 1999.

[6]   David M. Pozar, *Microwave Engineering,* Fourth Edition, John Wiley & Sons, New York, 2012

[7]   R. Ludwig, P. Bretchko, *RF Circuit Design,* Theory and Applications, Prentice Hall, Upper Saddle River, NJ, 2000

[8]   Jerry Sevick, *Transmission Line Transformers,* The American Radio Relay League, Newington, CT., 1990.

Problems

5-1.   For a load of 75 Ω with a 10 pF series capacitance we want to have maximum power transfer at 1 GHz. Find a source inductance that cancels the reactance of the load at this frequency.

5-2.   Analytically design all the L-networks that will match a complex load impedance, $Z_L = 15 - j10\ \Omega$, to a complex source impedance,

$Z_S = 25 + j20$ Ω, at a frequency of 1 GHz. Verify all the solutions by plotting the response of the matching networks.

5-3. Analytically design all the L-networks that will match a source impedance, $Z_S = 30$ Ω, to a complex load impedance, $Z_L = 25 + j20$ Ω, at a frequency of 1 GHz. Verify all the solutions by plotting the response of the matching networks.

5-4. Analytically design all the L-networks that will match a load impedance, $Z_L = 15$ Ω, to a source impedance, $Z_S = 75$ Ω, at a frequency of 1.5 GHz. Verify all the solutions by plotting the response of the matching networks.

5-5. The output impedance of a transmitter operating at a frequency of 2 GHz is: $Z_S = 30 + j10$ Ω. Analytically design all the matching L-networks such that the maximum power is delivered to the antenna whose input impedance is $Z_L = 10 + j15$ Ω.

5-6. Redesign the network of Problem 5-5 by matching the source and load impedances to a 50 Ω line. Compare fractional bandwidth and Q factors for examples 5.5 and 5.6.

5-7. Redesign the matching network in Example 5.5 with four equal-Q L-networks. Compare the Q and bandwidth of the two networks.

5-8 For a load and source resistor ratio of 5, find the minimum number of cascaded L sections needed to achieve a loaded Q of 0.5.

5-9. Use the Genesys Impedance Matching Utility to solve a complex load matching example with $R_L = 1000$ Ω in parallel with a 3 pF capacitance. For the source select $R_S = 50$ Ω, and for the load select the parallel RC with R = 1000 Ω and C = 3 pF. Compare the Pi and Tee Matching Network response.

(This page is intentionally left blank)

Chapter 6

Distributed Impedance Matching Networks

6.1 Introduction

Distributed networks are comprised of transmission line elements rather than discrete resistors, inductors, and capacitors. These transmission lines can take the form of the various transmission lines covered in chapter 2. At RF and microwave frequencies, where the wavelengths of the signals become comparable to the physical dimensions of the components, even lumped elements behave like distributed components. At microwave frequencies the distributed network is a more realizable form of a matching network than the lumped element networks.

6.2 Quarter-Wave Matching Network Design

In this subsection, based on the equations developed in section 6.2.1, two quarter-wave matching networks for $R_L = 2\ \Omega$ and $R_L = 150\ \Omega$ are designed. For each R_L the loaded Q factor and bandwidth is calculated at both 20 dB and 3 dB return loss.

Example 6-1 Design a quarter-wave network intended to match a 50 Ω source to a 2 Ω load at 100 MHz, as shown in Figure 6-1. Calculate the Q factor and the fractional bandwidths at 3 dB and 20 dB return loss. Compare the calculations with simulation in Genesys.

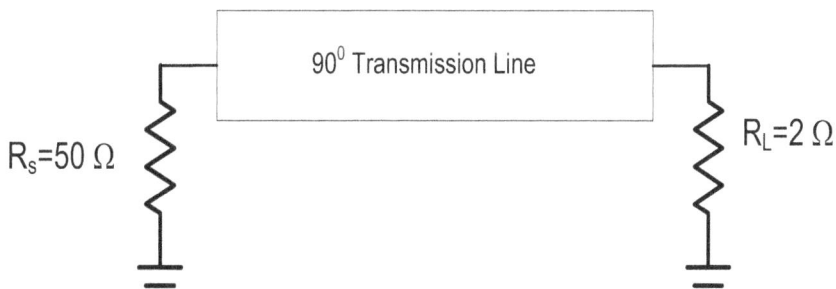

Figure 6-1 Matching quarter-wave transformer ($R_L < R_s$)

Solution: The characteristic impedance of the quarter wave matching network is:

$$Zo = \sqrt{(2)(50)} = 10\ \Omega$$

The schematic of quarter-wave matching network is shown in Figure 6-2.

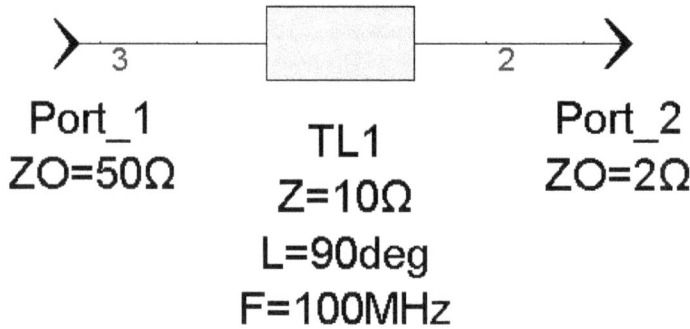

Port_1   TL1   Port_2
ZO=50Ω   Z=10Ω   ZO=2Ω
                          L=90deg
                          F=100MHz

Figure 6-2 Schematic of the quarter-wave matching network

Using the equations developed in the textbook, the characteristic impedance, loaded Q factor, and the bandwidths of the quarter wave matching network at 3 dB and 20 dB return loss are calculated in using MATLAB equations.

```
RS=50
RL=2
f=100e6
r=RL/RS
Z1=RS*sqrt(r)
FBW20dB=2-(4/PI)*acos((0.2*sqrt(r))/(sqrt(0.99)*(abs(1-r))))
BW20dB=(f*FBW20dB)
FBW3dB2=2-(4/PI)*acos((2*sqrt(r))/(abs(1-r)))
BW3dB2=(f*FBW3dB2)
Qe=1/FBW3dB2
```

Figure 6-3 Calculating Z1, Q factor, and fractional bandwidths ($R_L < R_S$)

The workspace variables are shown in Figure 6-4.

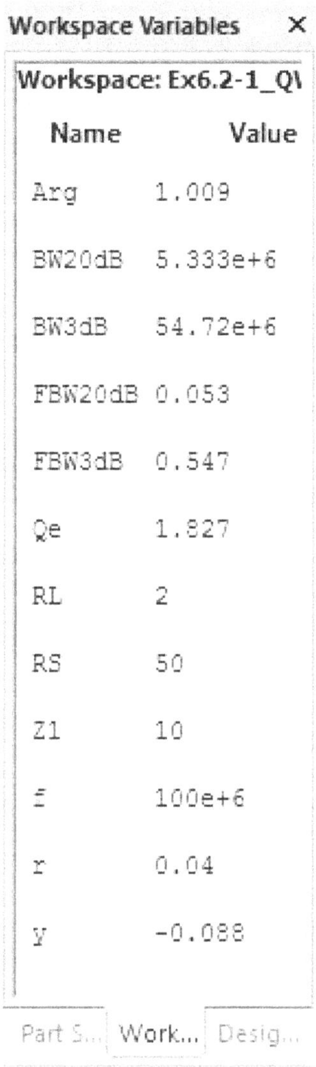

Figure 6-4: Workspace Variables

Viewing Workspace Variables show that the loaded Q factor is 1.827 and the bandwidths at 3dB and 20dB return loss are 54.72 MHz and 5.333 MHz, respectively. The fractional bandwidth at 20 dB return loss is only 5.333 % indicating a narrowband matching network. This is characteristic of a narrowband matching network with a load resistor that is much smaller than the source resistor. Next we analyze the matching network in Genesys to measure the same parameters. Set up a Linear Analysis to simulate the schematic of Figure 6-3. Display the input reflection coefficient, S11, and forward transmission, S21, from 50 MHz

to 150 MHz. Use markers to measure the simulated bandwidth at 3 dB
and 20 dB return loss as shown in Figure 6-5.

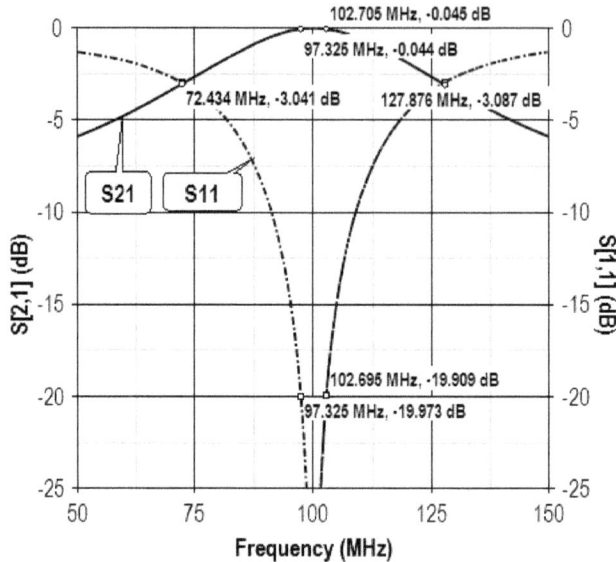

Figure 6-5 Response of the quarter-wave matching network ($R_L < R_S$)

Reading marker frequencies at 3 and 20 dB return loss, the corresponding
bandwidths are measured.

$$BW_{3dB} = 127.4 - 72.6 = 54.8 \text{ MHz}$$

$$BW_{20dB} = 102.7 - 97.3 = 5.4 \text{ MHz}$$

The loaded Q of the matching network is measured by using Equation (6-9),

$$Q_L = \frac{1}{FBW_{3dB}} = \frac{\sqrt{(127.4)(72.6)}}{127.4 - 72.6} = 1.75$$

Note that the measured Q factor and bandwidths are in close agreement
with the calculated values in Figure 6-4.

Example 6-2 Design a quarter-wave network intended to match a 50
Ohm source to a 150 Ω load. The design frequency is 100 MHz. Compare

the calculated Q factor and the fractional bandwidths, at 3 and 20 dB return loss, with the measurements.

Solution: The schematic of the quarter-wave matching network is shown in Figure 6-6.

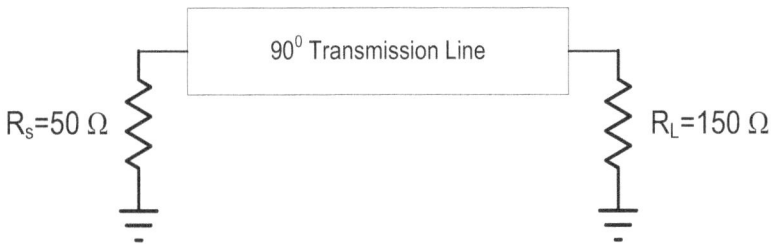

Figure 6-6 Quarter-wave matching network ($R_L > R_S$)

Using Equation (6-2) the characteristic impedance of the quarter-wave matching network is:

$$Z_1 = \sqrt{(50)(150)} = 86.6 \ \Omega$$

The Genesys schematic of the matching network is shown in Figure 6-7.

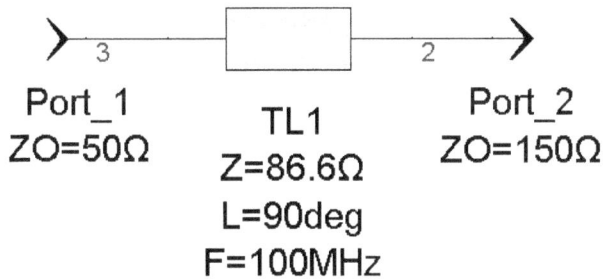

Figure 6-7 Schematic of the matching quarter-wave network

Using Equations developed in the textbook, the characteristic impedance, the loaded Q factor, and the bandwidths of the matching network at 3 and 20 dB return loss are calculated in Figure 6-8.

```
RS=50
RL=150
f=100e6
r=RL/RS
Z1=RS*sqrt(r)
FBW20dB=2-(4/PI)*acos((0.2*sqrt(r))/(sqrt(0.99)*(abs(1-r))))
BW20dB=(f*FBW20dB)
FBW3dB=2-(4/PI)*acos((2*sqrt(r))/(abs(1-r)))
BW3dB=(f*FBW3dB)
Qe=1/FBW3dB
```

Figure 6-8 Calculating Z1, bandwidths, and Q factor ($R_L > R_S$)

Viewing Workspace Variables show that the bandwidth at 20 dB return loss is 22.28 MHz. This is over four times the bandwidth that was achieved with the 2 Ω load impedance. Notice that the overall loaded Q of the network has significantly increased due to higher load impedance. Next analyze the matching network in Genesys by simulating the quarter wave transformer. Create a Linear Analysis and simulate the schematic of Figure 6-9. Display the input reflection coefficient, S11, and forward transmission, S21, from 50 to 150 MHz. Place markers at the 20 dB return loss as shown.

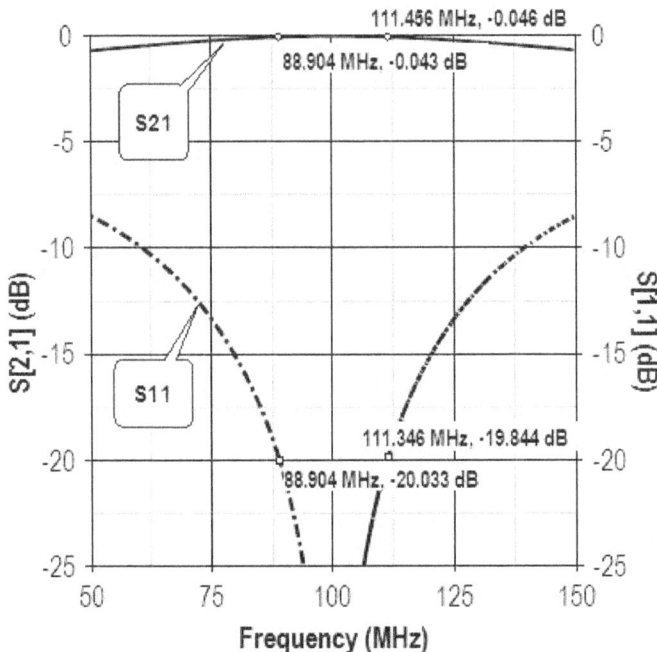

Figure 6-9 Response of the quarter-wave matching network

As Figure 6-9 shows the measured bandwidth at 20 dB return loss is:

$$111.3 - 88.9 = 22.4 \text{ MHz}$$

This agrees well with the result of calculation in the Equation Editor.

6.3 Quarter-Wave Matching Bandwidth

Equation (6-6) shows that the achievable bandwidth in a quarter-wave matching network is related to the ratio of the load to the source impedance (mismatch ratio) as well as the value of the input reflection coefficient. It is insightful to examine the relationship between these quantities when one of the parameters is swept in value.

6.4 Matching Bandwidth and Power Loss

The fractional bandwidth and power loss of a quarter-wave matching network can be calculated in an Equation Editor as a function of the input reflection coefficient, Γ_{IN}, and the normalized load resistor.

r. The procedure is listed here.

1. Enter equations for input reflection coefficient, power loss, and the conversion of reflection coefficient to return loss in dB.

2. Enter the desired values for the source and load resistors

3. Normalize the load resistor with respect to source resistor

4. Use Equation (6-6) to calculate the fractional bandwidth.

Example 6-3: For a 50 Ω source and 2 Ω load resistors, calculate the fractional bandwidth and power loss from $\Gamma = 0.1$ to $\Gamma = 0.707$.

Solution: The fractional bandwidth and power loss is calculated in the following Equation Editor.

```
vswr=Sweep1_Data.VSWR1
ReflCoef=(vswr-1)/(vswr+1)
Powerloss=(1-(1-(abs(ReflCoef)^2)))*100
RLdB=-20*log(abs(ReflCoef))
RS=50
RL=2
r=RL/RS
FBW=(2-(4/PI)*acos((2*(ReflCoef)*sqrt(r))/(sqrt((1-(ReflCoef)^2))*abs(1-r))))
```

Figure 6-10 Fractional bandwidth and power loss calculations for $R_L=2 \ \Omega$,

Sweep the parameters to display the fractional bandwidth and power loss for reflection coefficients from 0.1 to 0.707, as shown in Table 6-2.

	ReflCoef	FBW ...	RLdB	Powerloss
1	0.1	0.053	20	1
2	0.11	0.059	19.172	1.21
3	0.12	0.064	18.417	1.44
4	0.13	0.07	17.721	1.69
5	0.14	0.075	17.077	1.96
6	0.15	0.081	16.478	2.25
7	0.16	0.086	15.917	2.56
8	0.17	0.092	15.391	2.89
9	0.18	0.097	14.895	3.24
10	0.19	0.103	14.425	3.61
11	0.2	0.108	13.979	4
12	0.3	0.167	10.458	9
13	0.4	0.233	7.959	16
14	0.5	0.309	6.021	25
15	0.6	0.405	4.437	36
16	0.707	0.547	3.012	49.985

Table 6-2 Fractional bandwidth and power loss for $R_L=2 \ \Omega$

Table 6-2 shows that the fractional bandwidths at 0.1 reflection coefficient, corresponding to 20 dB return loss, is 5.3 % with 1% power loss while at 0.707 reflection coefficient, corresponding to 3 dB return loss, the fractional bandwidth is 54.7 % with 49.985 % power loss. Notice that the fractional bandwidths, at 3 and 20 dB return loss, are the same as calculated in Figure 6-4.

Example 6-4: For a 50 Ωsource and 150 Ω load resistors, calculate the fractional bandwidth and power loss from$\Gamma = 0.1$ to $\Gamma = 0.707$.

Solution: Following the procedure, the calculations are shown in Figure 6-11. Sweep the parameters and display the result, as shown in Table 6-3.

```
vswr=Sweep1_Data.VSWR1
ReflCoef=(vswr-1)/(vswr+1)
Powerloss=(1-(1-(abs(ReflCoef)^2)))*100
RLdB=-20*log(abs(ReflCoef))
RL=150
r=RL/RS
FBW=(2-(4/PI)*acos((2*(ReflCoef)*sqrt(r))/(sqrt((1-(ReflCoef)^2))*abs(1-r))))
```

Figure 6-11 Fractional bandwidth and power loss calculation, $R_L = 150\ \Omega$

	ReflCoef	FBW (rad)	RLdB	Powerloss
1	0.1	0.223	20	1
2	0.11	0.246	19.172	1.21
3	0.12	0.269	18.417	1.44
4	0.13	0.292	17.721	1.69
5	0.14	0.315	17.077	1.96
6	0.15	0.339	16.478	2.25
7	0.16	0.362	15.917	2.56
8	0.17	0.386	15.391	2.89
9	0.18	0.411	14.895	3.24
10	0.19	0.435	14.425	3.61
11	0.2	0.46	13.979	4
12	0.3	0.733	10.458	9
13	0.4	1.091	7.959	16
14	0.5	2	6.021	25
15	0.6	1.≠QO	4.437	36
16	0.707	1.≠QO	3.012	49.985

Table 6-3 Fractional bandwidth and power loss measurements for R_s=50 Ω and R_L=150 Ω

Table 6-3 shows that the fractional bandwidth at $\Gamma = 0.1$, corresponding to 20 dB return loss, is 22.3 % with a 1% power loss. The fractional bandwidth is the same as calculated in Figure 6-8. The higher fractional bandwidth in this example, compared to its value in Example 6-3, is due to the lower ratio of the load to source resistor.

6.5 Single-Stub Matching Network Design

Example 6-5: Design a single-stub network to match a load resistance $Z_L = 2 - j5$ Ω to a resistive source $R_S = 50$ Ω at 100 MHz. Display the response and measure the fractional bandwidth at 3 dB and 20 dB return loss.

Solution: The design example is shown in Figure 6-12.

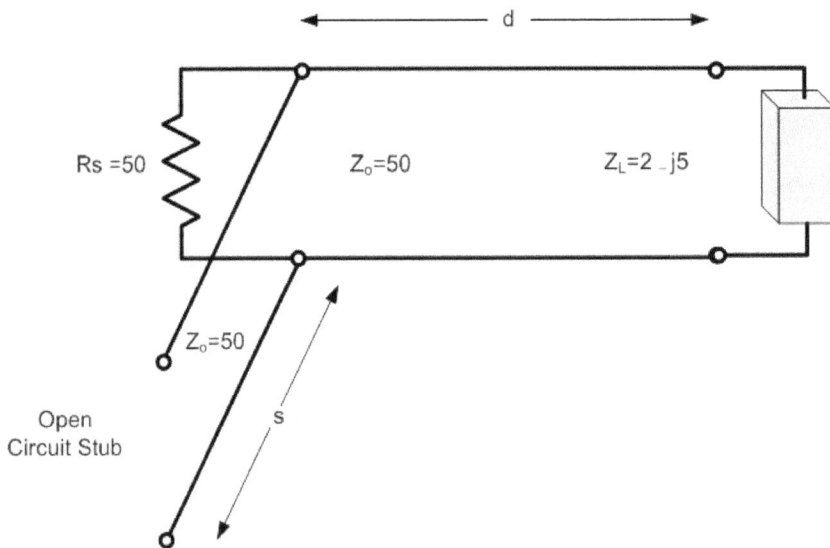

Figure 6-12 Complex load to resistive source matching network

Equation Editor of Figure 6-13 calculates the electrical length of the line and stub matching network.

```
RS=50
RL=2
XL=-5
f=100e6
r=RL/RS
x=XL/RS
t1=(x+sqrt(r*(r^2+x^2-2*r+1)))/(r-1)
t2=(x-sqrt(r*(r^2+x^2-2*r+1)))/(r-1)
d1=360*(atan(t1))/(2*pi)
d2=360*(atan(t2))/(2*pi)
B1=(x*t1^2+(r^2+x^2-1)*t1-x)/(RS*(r^2+x^2+t1^2+2*x*t1))
B2=(x*t2^2+(r^2+x^2-1)*t2-x)/(RS*(r^2+x^2+t2^2+2*x*t2))
so1=360*(pi-atan(RS*B1))/(2*pi)
so2=-360*atan(RS*B2)/(2*pi)
```

Figure 6-13 Calculation of line and stub lengths

The calculations in Figure 6-13 show that the problem has two solutions. Either solution can be used in the design of the single-stub matching network. For both line and stub usually the shorter line lengths are chosen. For this example we select the shorter electrical lengths in the second solution; d2 for the line and so2 for the open-circuited stub. The Genesys schematic of the single-stub matching network is shown in Figure 6-14. The simulated response of Figure 6-15 shows that the 3-dB bandwidth is about 11 MHz while the 20-dB bandwidth is about 1 MHz.

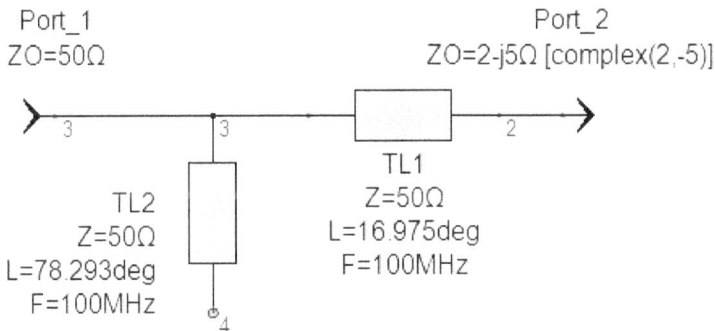

Port_1
ZO=50Ω

Port_2
ZO=2-j5Ω [complex(2,-5)]

TL1
Z=50Ω
L=16.975deg
F=100MHz

TL2
Z=50Ω
L=78.293deg
F=100MHz

Figure 6-14 Schematic of the single-stub matching network

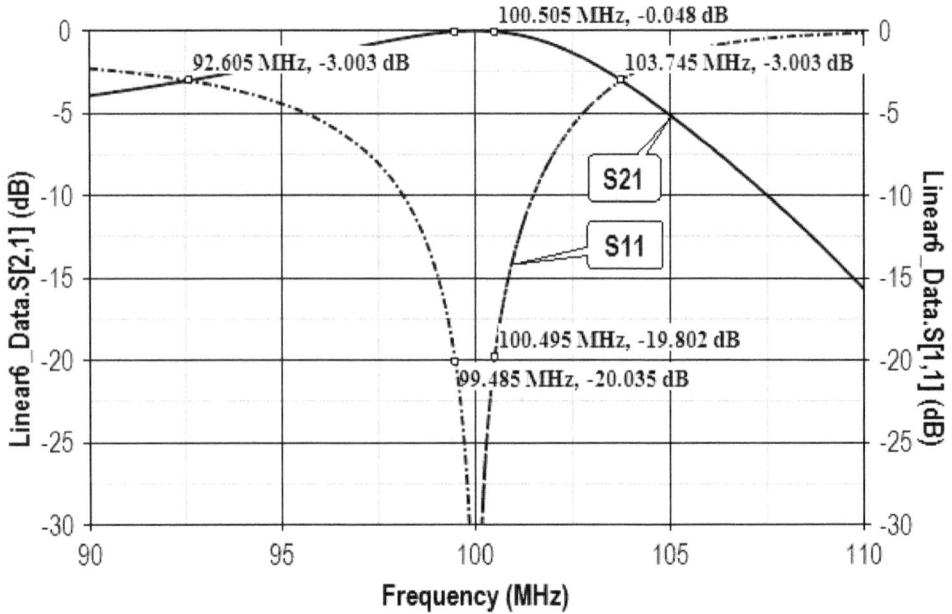

Figure 6-15 Response of the single-stub matching network

6.6 Graphical Design of Single-Stub Networks

In this section a 50 Ω source will be matched to a 5 - 25j Ω load at a frequency of 1000 MHz. Both open-circuited and short-circuited shunt stubs will be considered.

Design a single-stub network to match the Z_L = 5 - j25 Ω load impedance to 50 Ohm source resistor using an open-circuited stub.

Create a new workspace in ADS and open a new schematic window. Insert the S_Params Template and add a series transmission line (TLIN) between the source and load terminations. Wire up the components and set the component values as shown in Figure 6-15.

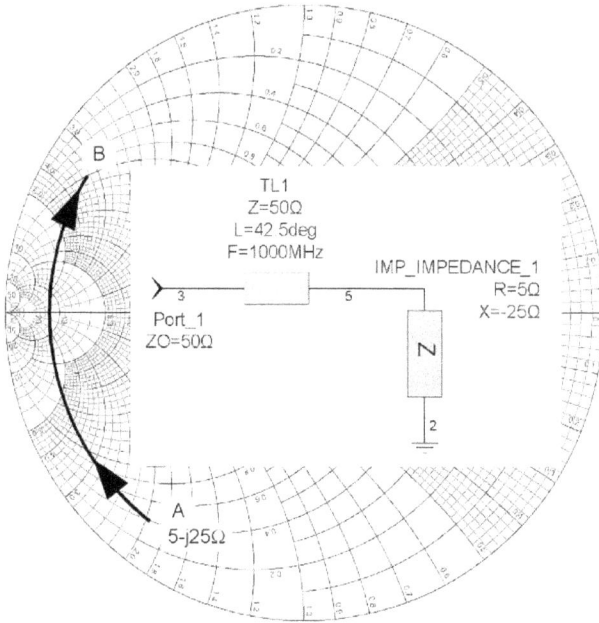

Figure 6-16 Adding series transmission line (electrical length=42.5°)

Then add an open-circuited shunt transmission line and tune the length until the impedance moves to the center of the chart (50Ω). As the schematic of Figure 6-17 shows, a 73° length of transmission line would be required.

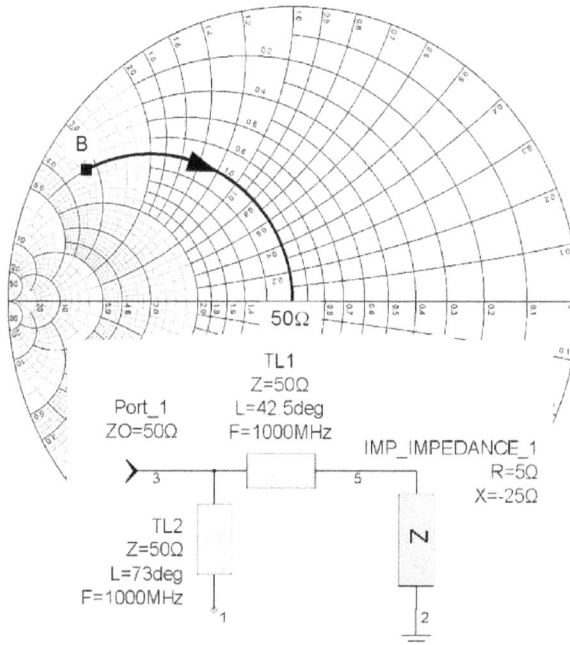

Figure 6-17 Adding open-circuited shunt transmission line (73°) to network

6.7 Graphical Design Using Short Circuit Stubs

In this Section a short-circuited shunt stub will be used.

Design a single-stub network to match the 5 - j25 Ωload impedance to 50 Ohm source resistor using short-circuited stub.

A short-circuited shunt transmission line can be used to perform the function of the shunt stub. Using a short-circuited shunt transmission line, the series transmission line should intersect the unit conductance circle on the bottom half of the Smith Chart. Add a series transmission line of 11° electrical length to move the impedance at point A to intersect the unit conductance circle at point B as shown in Figure 6-26. Then add the short-circuited shunt transmission line as shown in Figure 6-18. Tune the length to 17° to move the impedance to the center of the Smith Chart.

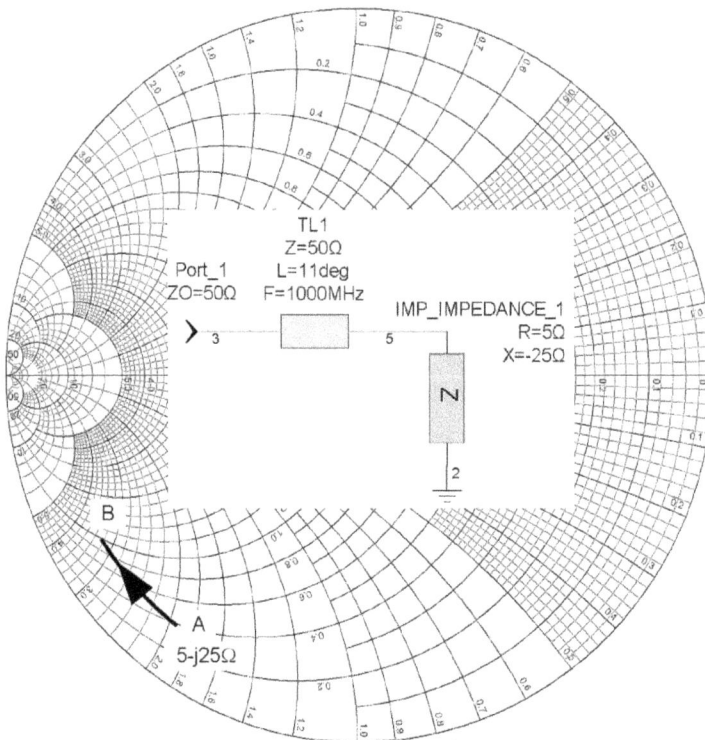

Figure 6-18 Adding series transmission line (11°) to 5-j25 Ω

To move point B to the center of Smith chart, add a short-circuited shunt transmission line and tune the length of the line until the impedance moves to the center of the chart (50Ω). As the schematic of Figure 6-19 shows, a 17° length of transmission line would be required.

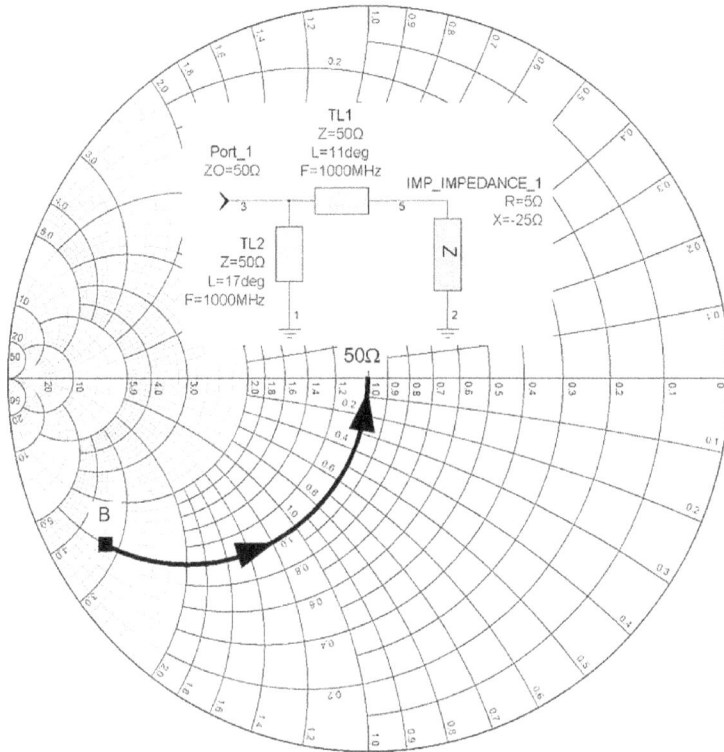

Figure 6-19 Adding short-circuited shunt transmission line (17°)

6.8 Design of the Cascaded Matching Networks

When the impedance of the load and the source are both complex, we can define a virtual resistor and design single-stub networks to match the complex impedances to the virtual resistor. The final matching network is obtained by cascading the two single-stub matching networks. The procedure is demonstrated in the following example.

Example 6-6: Design cascaded single-stub networks to match a complex load $Z_L = 10 - j5 \ \Omega$ to a complex source $Z_S = 50 - j15 \ \Omega$ at 100 MHz. Calculate the electrical lengths of the lines and the fractional bandwidth at 3 dB and 20 dB return loss. Display the simulated response and verify the calculations with simulation in Genesys.

Solution: The following procedure is used to match the two complex impedances

1. Find the intermediate resistor, $R = \sqrt{R_L R_S}$

2. Design a single-stub matching network between *R1* and the source impedance
3. Design a second matching network between *R1* and the load impedance
4. Cascade the two single-stub networks

First Solution: For this example we match the source to R_1, where: $R_1 = \sqrt{(10)(50)} = 22.36\,\Omega$. There are two matching networks in each case.

```
RS=22.36
RL=50
XL=-15
f=100e6
r=RL/RS
x=XL/RS
t1=(x+sqrt(r*(r^2+x^2-2*r+1)))/(r-1)
t2=(x-sqrt(r*(r^2+x^2-2*r+1)))/(r-1)
d1=360*(atan(t1))/(2*pi)
d2=360*(pi+atan(t2))/(2*pi)
B1=(x*t1^2+(r^2+x^2-1)*t1-x)/(RS*(r^2+x^2+t1^2+2*x*t1))
B2=(x*t2^2+(r^2+x^2-1)*t2-x)/(RS*(r^2+x^2+t2^2+2*x*t2))
sol=360*(pi-atan(RS*B1))/(2*pi)
so2=-360*atan(RS*B2)/(2*pi)
```

Figure 6-20 Calculation of the first line and stub matching network

For the first solution we select, from the Workspace Variables, d2=114.019 and so2=43.245 as shown in Figure 6-29.

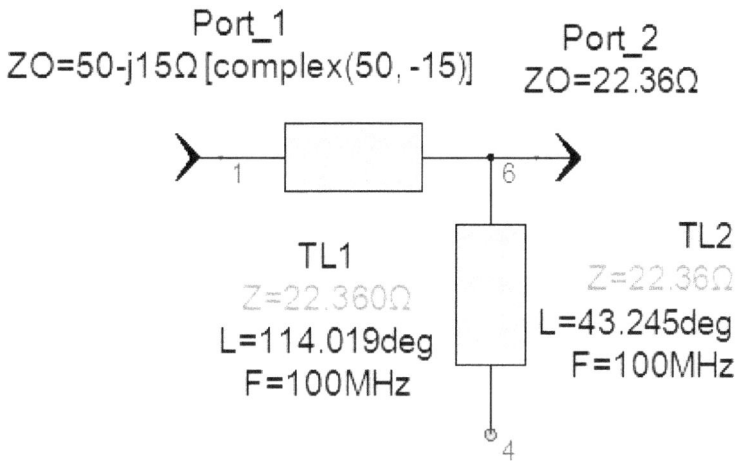

Figure 6-21 Schematic of the first single-stub matching network

Example 6-7(Second solution): Design single-stub networks to match a complex load $Z_L = 10 - j5\ \Omega$ to a complex source $Z_S = 50 - j15\ \Omega$ at 100 MHz. Calculate the electrical lengths of the lines and the fractional bandwidths of the matching network at 3 dB and 20 dB return loss.

Second Solution: Design of the second matching network is shown in Figure 6-22.

```
RS1=22.36
RL1=10
XL1=-5
f1=100e6
r1=RL1/RS1
x1=XL1/RS1
t11=(x1+sqrt(r1*(r1^2+x1^2-2*r1+1)))/(r1-1)
t21=(x1-sqrt(r1*(r1^2+x1^2-2*r1+1)))/(r1-1)
d11=360*(pi+atan(t11))/(2*pi)
d21=360*(atan(t21))/(2*pi)
B11=(x1*t11^2+(r1^2+x1^2-1)*t11-x1)/(RS1*(r1^2+x1^2+t11^2+2*x1*t11))
B21=(x1*t21^2+(r1^2+x1^2-1)*t21-x1)/(RS1*(r1^2+x1^2+t21^2+2*x1*t21))
so11=360*(pi-atan(RS1*B11))/(2*pi)
so21=-360*atan(RS1*B21)/(2*pi)
```

Figure 6-22 Calculation of the second single-stub matching network

For the second solution we select from the Workspace Variables d2=48.39 and so2=41.722 degrees as shown in Figure 6-23.

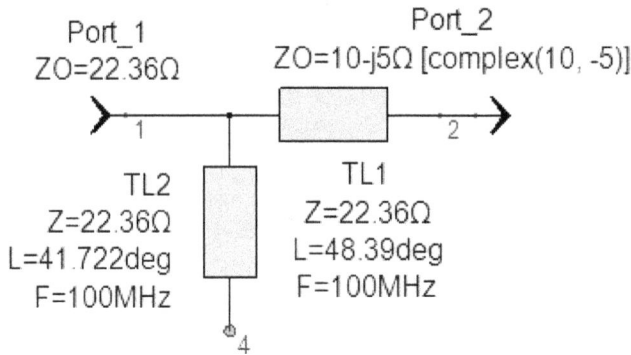

Figure 6-23 Schematic of the second single-stub matching network

Now connect both line and stub matching networks in cascade to obtain the complete matching network, as shown in Figure 6-24.

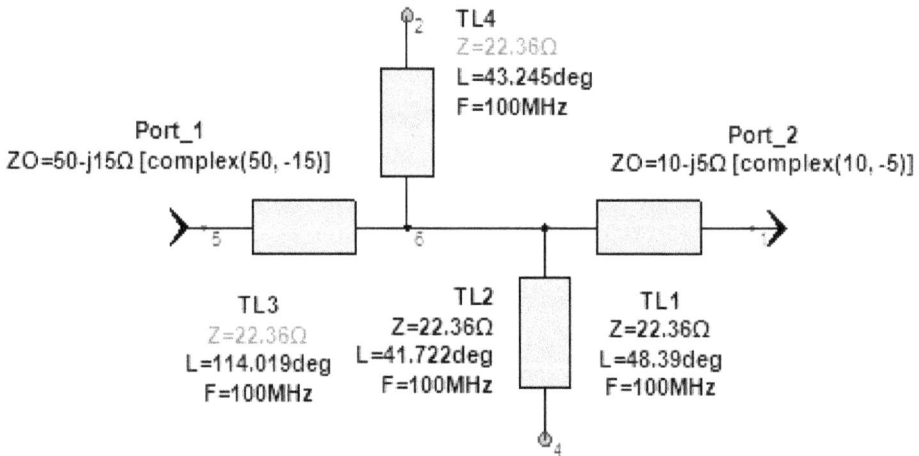

Figure 6-24 Cascading single-stub matching networks

The simulated response of the complete network is shown in Figure 6-25. Notice that the cascaded single-stub matching network has about 100 % fractional bandwidth at the 3 dB return loss and 47% fractional bandwidth at the 12 dB return loss. The matching network response is shown in Figure 6-25.

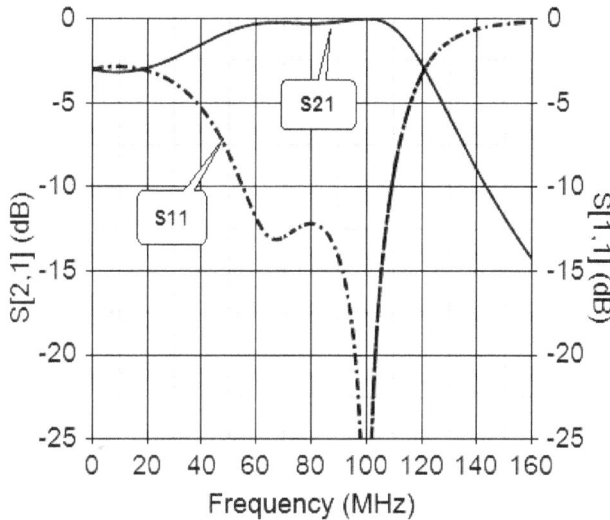

Figure 6-25 Response of the cascaded matching network

6.9 Broadband Quarter-Wave Matching Networks

In Examples 6-1 and 6-2, the bandwidth of a single quarter-wave transformer matching network with a large reflection coefficient is less than 10 % which is considered to be a narrowband matching network. We can increase the bandwidth by cascading two or more quarter-wave transformers to achieve a broadband matching network. To analytically design a broadband matching network with N quarter-wave transformers first use the Equation Editor to calculate the characteristic impedance of each quarter-wave transformer then cascade all the sections into one matching network. Let Rs and R_L be the source and load impedances to be matched by the N quarter-wave transformation network. The characteristic impedance of each section can be calculated from the following Equation.

$$Z_n = R_S(r)^{(2n-1)/2N} \qquad n = 1, 2, \ldots, N$$

Where $r = R_L/R_S$ is the normalized load resistance and N is the number of quarter-wave transformers.

Example, 6-8: Design a three-section quarter-wave network to match a load resistance $R_L = 2 \ \Omega$ to a resistive source $R_S = 50 \ \Omega$ at 100 MHz.

(a) Calculate the characteristic impedance of the quarter-wave transformers, the Q factor and the fractional bandwidths of thematching network at 3 and 20 dB return loss

(b) Display the simulated response and compare the measurements with themeasurements of single quarter-wave transformer matching network.

Solution: (a) Using the following Equations, the characteristic impedances of the 3-section quarter-wave transformers are:

$$Z1 = R_S(r)^{1/2N} = 50(0.04)^{1/6} = 29.24$$

$$Z2 = R_S(r)^{3/2N} = 50(0.04)^{3/6} = 10.00$$

$$Z3 = R_S(r)^{5/2N} = 50(0.04)^{5/6} = 3.42$$

The intermediate resistors are obtained from the following Equation.

$$R_n = R_S(r)^{\frac{n}{N}} \quad n = 1, 2, 3, \ldots, N\text{-}1$$

Therefore, the two intermediate resistors are:

$$R_1 = 50(0.04)^{\frac{1}{3}} = 17.1 \ \Omega$$

$$R_2 = 50(0.04)^{\frac{2}{3}} = 5.848 \ \Omega$$

These calculations are solved in the MATLAB Equation of Figure 6-26. The Genesys schematic of the three-section quarter-wave matching network is shown in Figure 6-27.

```
RS3=50
RL3=2
f3=100e6
r3=RL3/RS3
N=3


Z1=RS3*r3^(1/(2*N))
Z2=RS3*r3^(3/(2*N))
Z3=RS3*r3^(5/(2*N))
R1=RS3*r3^(1/N)
R2=RS3*r3^(2/N)
```

Figure 6-26 Calculation of characteristic impedances

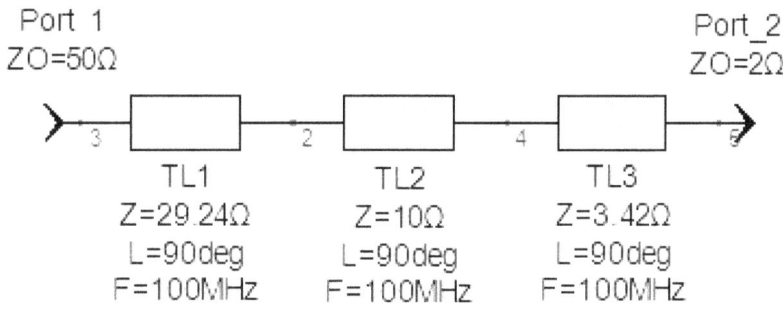

Figure 6-27 Three-section quarter-wave matching network

The simulated response of the matching network is shown in Figure 6-28.

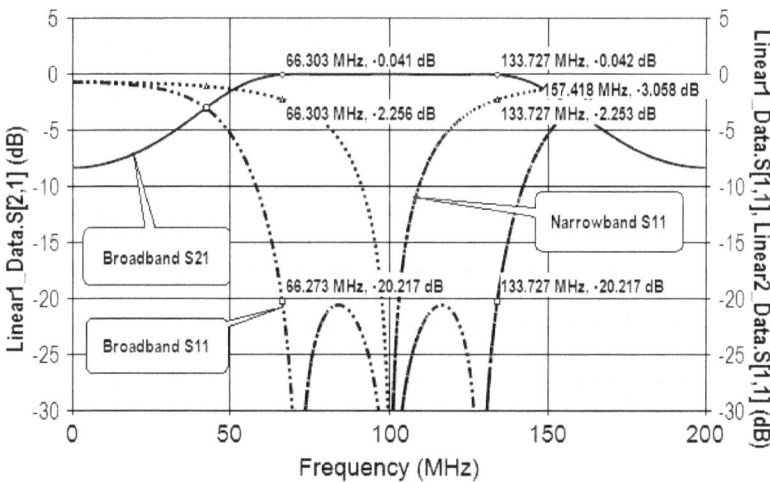

Figure 6-28 Response of the cascaded matching network

Figure 6-28 shows that the bandwidth simulated at the 3 dB return loss is:

$$BW_{3dB} = 157.6 - 42.4 = 115.2 \text{ MHz}$$

$$FBW_{3dB} = \frac{115.2x100}{\sqrt{(157.6).(42.4)}} = 140.8\%$$

Similarly the simulated bandwidth at the 20 dB return loss is:

$$BW_{20dB} = 133.8 - 66.2 = 67.6 \quad \text{MHz}$$

$$FBW_{20dB} = \frac{67.6x100}{\sqrt{(133.8).(66.2)}} = 71.8\%$$

The measured Q factor for the matching network is,

$$Q = \frac{1}{1.408} = 0.71$$

The wide bandwidth and lower Q factor is an indication that by adding two more quarter-wave sections the Q factor has reduced to less than one half and the 3-dB bandwidth has more than doubled compared to a single quarter-wave matching network.

Example 6-9: Design a broadband quarter-wave network to match a complex load, $Z_L = 10 - j5 \ \Omega$ to $50 \ \Omega$ source impedance at 100 MHz. Calculate the bandwidth at 20 dB return loss and compare with the bandwidth of a single-stub matching network.

Solution: To design a broadband matching network between a resistive source and complex load impedance; first design a 5 section matching network between the real source and the real part of the load impedance. Then replace the quarter-wave transformer adjacent to the load with a single-stub that matches the complex load to the real resistor. Calculation of the broadband matching network between R_L and R_S is shown in the Equation Editor of Figure 6-29. The Genesys schematic of the five-section matching network is shown in Figure 6-30.

$$RS=50$$
$$RL=10$$
$$f=100e6$$
$$r=RL/RS$$
$$N=5$$
$$Z1=RS*r^{(1/(2*N))}$$
$$Z2=RS*r^{(3/(2*N))}$$
$$Z3=RS*r^{(5/(2*N))}$$
$$Z4=RS*r^{(7/(2*N))}$$
$$Z5=RS*r^{(9/(2*N))}$$
$$R1=RS*r^{(1/N)}$$
$$R2=RS*r^{(2/N)}$$
$$R3=RS*r^{(3/N)}$$
$$R4=RS*r^{(4/N)}$$

Figure 6-29 Calculating 4 resistors and 5 characteristic impedances

The five section quarter-wave schematic is shown in Figure 6-30.

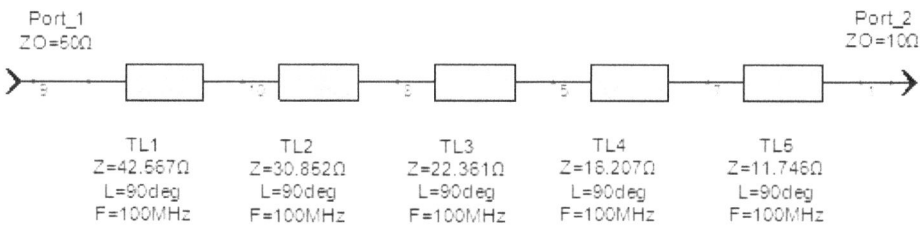

Port_1
ZO=50Ω

Port_2
ZO=10Ω

TL1	TL2	TL3	TL4	TL5
Z=42.567Ω	Z=30.852Ω	Z=22.361Ω	Z=16.207Ω	Z=11.746Ω
L=90deg	L=90deg	L=90deg	L=90deg	L=90deg
F=100MHz	F=100MHz	F=100MHz	F=100MHz	F=100MHz

Figure 6-30 Five-section quarter-wave transformer matching network

Next replace the quarter-wave section adjacent to the load with a single-stub matching network. The design of the single-stub matching network is shown in the following Equation Editor. Note that the source resistor, R4 = 13.797 Ω, was calculated in the Equation Editor of Figure 6-31. The single-stub matching network is shown in Figure 6-32.

```
RS=13.797
RL=10
XL=-5
f=100e6
r=RL/RS
x=XL/RS
t1=(x+sqrt(r*(r^2+x^2-2*r+1)))/(r-1)
t2=(x-sqrt(r*(r^2+x^2-2*r+1)))/(r-1)
d1=360*(pi+atan(t1))/(2*pi)
d2=360*(atan(t2))/(2*pi)
B1=(x*t1^2+(r^2+x^2-1)*t1-x)/(RS*(r^2+x^2+t1^2+2*x*t1))
B2=(x*t2^2+(r^2+x^2-1)*t2-x)/(RS*(r^2+x^2+t2^2+2*x*t2))
so1=360*(pi-atan(RS*B1))/(2*pi)
so2=-360*atan(RS*B2)/(2*pi)
```

Figure 6-31 Second line and stub calculations

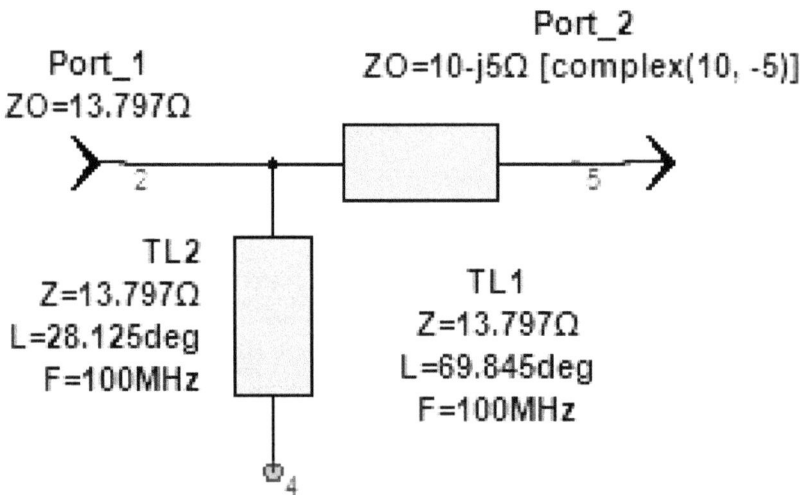

Figure 6-32 Schematic of the single-stub matching network

Now cascade the line and stub with four quarter-wave networks to form the final design of the matching network in Figure 6-33. The simulated response of the matching network is shown in Figure 6-34.

Port_1
ZO=50Ω

Port_2
ZO=10-j5Ω[complex(10, -5)]

TL4
Z=42.567Ω
L=90deg
F=100MHz

TL3
Z=30.852Ω
L=90deg
F=100MHz

TL5
Z=22.361Ω
L=90deg
F=100MHz

TL8
Z=16.207Ω
L=90deg
F=100MHz

TL2
Z=13.797Ω
L=28.125deg
F=100MHz

TL1
Z=13.797Ω
L=69.845deg
F=100MHz

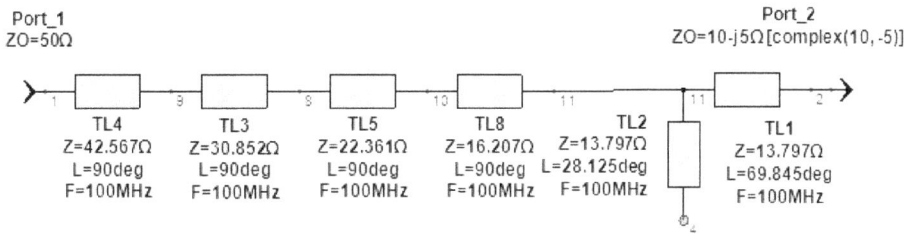

Figure 6-33 Schematic of the broadband matching network

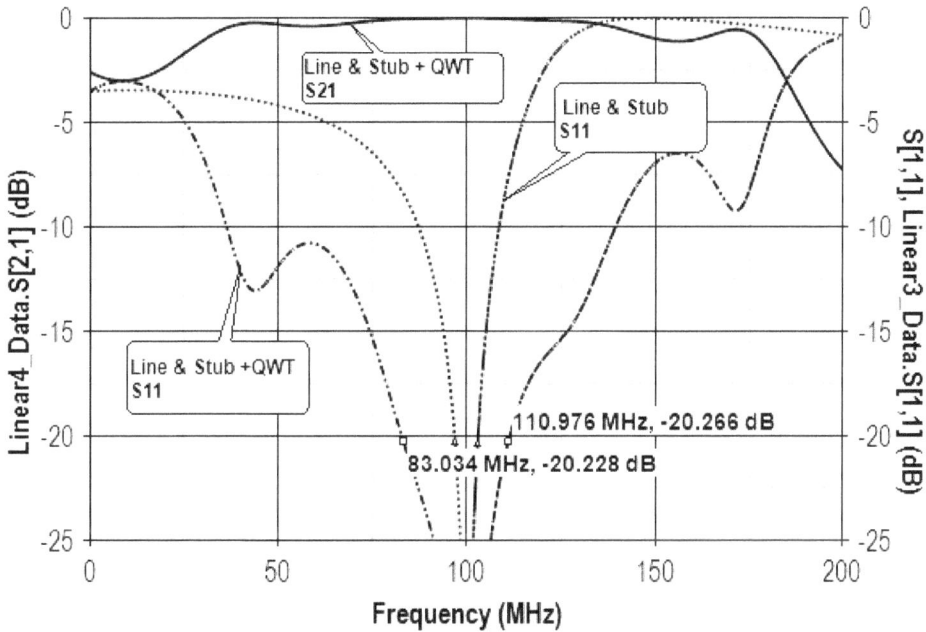

Figure 6-34 Response of the broadband matching network

Figure 6-34 shows that the bandwidth at 20 dB return loss is:

$$BW_{20dB} = 110 - 83 = 27 \text{ MHz}$$

And the fractional bandwidth is:

$$FBW_{20dB} = \frac{27}{\sqrt{(110).(83)}} = 28.2 \%$$

The same numbers for the single-stub matching network are:

$$BW_{20dB} = 102.8 - 96.8 = 6 \quad MHz$$

$$FBW_{20dB} = \frac{6x100}{\sqrt{(102.8).(96.8)}} = 6\%$$

Notice that fractional bandwidth for the broadband at 20 dB return loss is 28.2 % as opposed to only 6 % for the narrowband single- stub matching network. Therefore, we have increased the matching bandwidth at 20 dB return loss by more than four times over the single-stub matching network.

Example 6-10: Design a broadband quarter-wave network to match a complex load, $Z_L = 150 - j30$ to 50 Ω source impedance at 100 MHz. Display the simulated response and measure the bandwidth at 20 dB return loss. Compare the results with the singe-stub matching network.

Solution: Use the same method as in Example 6.6.1 to design the broadband matching network. The design of the broadband quarter-wave transformer matching network with five quarter-wave sections is shown in Figure 6-35.

```
RS=50
RL=150
f=100e6
r=RL/RS
N=5

Z1=RS*r^(1/(2*N))
Z2=RS*r^(3/(2*N))
Z3=RS*r^(5/(2*N))
Z4=RS*r^(7/(2*N))
Z5=RS*r^(9/(2*N))

R1=RS*r^(1/N)
R2=RS*r^(2/N)
R3=RS*r^(3/N)
R4=RS*r^(4/N)
```

Figure 6-35 Calculating 5 characteristic impedances and 4 intermediate resistors

The Genesys schematic of the five-section quarter-wave matching network is shown in Figure 6-36. The simulated response of the quarter-wave matching network is shown in Figure 6-37.

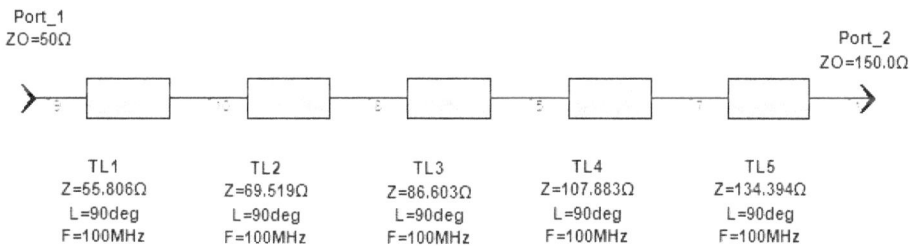

Port_1
ZO=50Ω

Port_2
ZO=150.0Ω

TL1	TL2	TL3	TL4	TL5
Z=55.806Ω	Z=69.519Ω	Z=86.603Ω	Z=107.883Ω	Z=134.394Ω
L=90deg	L=90deg	L=90deg	L=90deg	L=90deg
F=100MHz	F=100MHz	F=100MHz	F=100MHz	F=100MHz

Figure 6-36 Five-section quarter-wave matching network

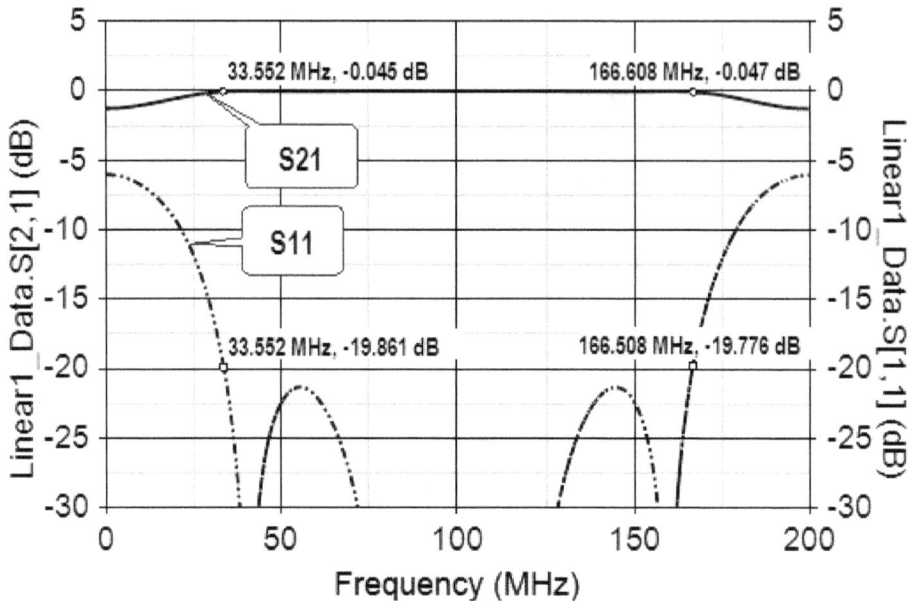

Figure 6-37 Response of the matching network

Next replace the quarter-wave transformer adjacent to the load with a single-stub matching network. The design of the single-stub matching network is shown in the following Equation Editor. Note that the source resistor, R4 = 13.797 Ω, was calculated in the Equation Editor of Figure 6-35.

```
RS=120.411
RL=150
XL=-30
f=100e6
r=RL/RS
x=XL/RS
t1=(x+sqrt(r*(r^2+x^2-2*r+1)))/(r-1)
t2=(x-sqrt(r*(r^2+x^2-2*r+1)))/(r-1)
d1=360*(atan(t1))/(2*pi)
d2=360*(pi+atan(t2))/(2*pi)
B1=(x*t1^2+(r^2+x^2-1)*t1-x)/(RS*(r^2+x^2+t1^2+2*x*t1))
B2=(x*t2^2+(r^2+x^2-1)*t2-x)/(RS*(r^2+x^2+t2^2+2*x*t2))
so1d=360*(pi-atan(RS*B1))/(2*pi)
so2d=-360*atan(RS*B2)/(2*pi)
```

Figure 6-38 Equation Editor showing the second line and stub calculations

The single-stub matching network is shown in Figure 6-39.

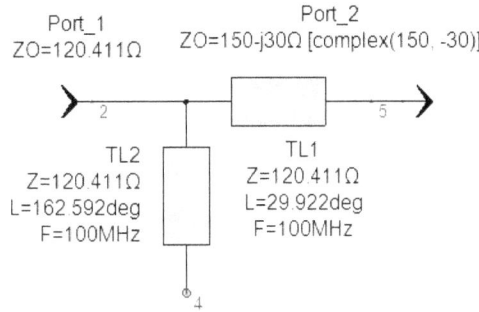

Port_1
ZO=120.411Ω

Port_2
ZO=150-j30Ω [complex(150, -30)]

TL2
Z=120.411Ω
L=162.592deg
F=100MHz

TL1
Z=120.411Ω
L=29.922deg
F=100MHz

Figure 6-39 Schematic of the single-stub matching network

Now cascade the two matching networks to form the final design of broadband matching network, as shown in Figure 6-40. The simulated response of the quarter-wave matching network is shown in Figure 6-41.

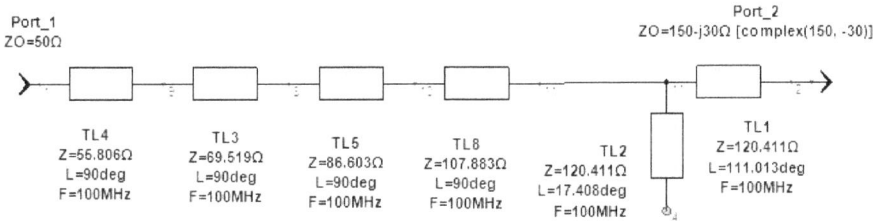

Port_1
ZO=50Ω

Port_2
ZO=150-j30Ω [complex(150, -30)]

TL4
Z=55.806Ω
L=90deg
F=100MHz

TL3
Z=69.519Ω
L=90deg
F=100MHz

TL5
Z=86.603Ω
L=90deg
F=100MHz

TL8
Z=107.883Ω
L=90deg
F=100MHz

TL2
Z=120.411Ω
L=17.408deg
F=100MHz

TL1
Z=120.411Ω
L=111.013deg
F=100MHz

Figure 6-40 Schematic of the broadband matching network

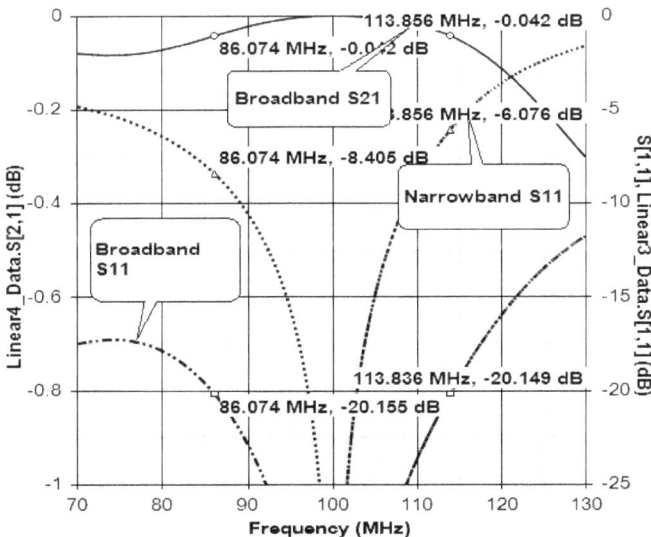

Figure 6-41 Simulated response of the broadband matching network

Figure 6-41 shows that the bandwidths of the broadband matching network at 20 dB return loss are:

$$BW_{20dB} = 113.8 - 86.0 = 27.8 \text{ MHz}$$

$$FBW_{20dB} = \frac{27.8 x 100}{\sqrt{(113.8).(86.0)}} = 28.1 \%$$

The same numbers for the narrowband single-stub matching network are:

$$BW_{20dB} = 102.8 - 97.1 = 5.7 \text{ MHz}$$

$$FBW_{20dB} = \frac{5.7 x 100}{\sqrt{(102.8).(97.1)}} = 5.7 \%$$

Notice that the fractional bandwidth at 20 dB return loss is about 28.1 % as opposed to only 5.7 % for the narrowband matching network. This is an indication that, by adding four quarter-wave sections to the single-stub matching network, we have increased the matching bandwidth at 20 dB return loss nearly five times over a single-stub matching network.

References and Further Readings

[1]   Keysight Technologies, *Genesys 2015.08, Users Guide*, www.keysight.com

[2]   Guillermo Gonzales, *Microwave Transistor Amplifiers – Analysis and Design*, Second Edition, Prentice Hall Inc., Upper Saddle River, NJ.

[3]   Randy Rhea, *The Yin-Yang of Matching: Part 1 – Basic Matching Concepts*, High Frequency Electronics, March 2006

[4]   Steve C. Cripps, *RF Power Amplifiers for Wireless Communications*, Artech House Publishers, Norwood, MA. 1999

[5]   David M. Pozar, *Microwave Engineering*, Fourth Edition, John Wiley & Sons, New York, 2012

[6]   Ali A. Behagi and Stephen D. Turner, *Microwave and RF Engineering,* A Simulation Approach with Keysight Genesys Software, BT Microwave LLC, March, 2015

Problems

6-1.   Design a quarter-wave transmission line to match a load resistance $R_L = 5$ Ω to a resistive source $R_S = 75$ Ω at 500 MHz. Calculate the characteristic impedance of the quarter-wave line, the Q factor and the fractional bandwidths of the matching network at 3 and 20 dB return loss. Display the simulated response and compare the calculations with measurements. For the quarter-wave matching network, calculate the fractional bandwidth and power loss from $\Gamma = 0.1$ to $\Gamma = 0.707$.

6-2.   Design a quarter-wave transmission line to match a load resistance $R_L = 100$ Ω to a resistive source $R_S = 25$ Ω at 600 MHz. Calculate the characteristic impedance of the quarter-wave line, the Q factor and the fractional bandwidths of the matching network at 3 dB and 20 dB return loss. Display the simulated response and compare the calculations with measurements. For the quarter-wave matching network, calculate the fractional bandwidth and power loss from $\Gamma = 0.1$ to $\Gamma = 0.707$.

6-3.   Design a single-stub network to match a load resistance $Z_L = 5 - j5$ Ω to a resistive source $R_S = 75$ Ω at 700 MHz. Calculate the electrical lengths of the matching line and stub and the fractional bandwidths of the matching network at 3 and 20 dB return loss. Display the simulated response and compare the calculations with measurements.

6-4.   Design a single-stub network to match a load resistance Z_L = 100 + j20 Ω to a resistive source R_S = 40 Ω at 800 MHz. Calculate the electrical lengths of the matching line and stub and the fractional bandwidths of the matching network at 3 dB and 20 dB return loss. Display the simulated response and compare the calculations with measurements.

6-5.   Design a single-stub network to match a complex load Z_L = 20 + j5 Ω to a complex source Z_S = 50 + j20 Ω at 1000 MHz. Calculate the electrical lengths of the matching line and stub and the fractional bandwidths of the matching network at 3 dB and 20 dB return loss. Display the simulated response and verify the calculations with measurements.

6-6.   Design a three-section quarter-wave network to match a load resistance R_L = 5 Ω to a resistive source R_S = 50 Ω at 100 MHz. Calculate the characteristic impedance of the quarter-wave line, the Q factor and the fractional bandwidths of the matching network at 3 and 20 dB return loss. Display the simulated response and compare the measurements with the singe quarter-wave matching network.

6-7.   Design a three-section quarter-wave network to match a load resistance R_L = 100 Ω to a resistive source R_S = 25 Ω at 900 MHz. Calculate the fractional bandwidths of the matching network at 3 and 20 dB return loss. Display the simulated response and compare the measurements with the singe quarter-wave matching network.

6-8.   Design a broadband network to match a complex load, Z_L = 25 + j10 Ω to 75 Ω source impedance at 300 MHz. Measure the bandwidth at 20 dB return loss and compare it with the results of singe-stub matching network.

Chapter 7

Single Stage Amplifier Design

7.1 Introduction

In this chapter we design four single stage amplifiers that require four different impedance matching techniques.

1. Maximum Gain Amplifier Design

2. Specific Gain Amplifier Design

3. Low Noise Figure Amplifier Design

4. Power Amplifier Design

Most amplifier designs actually involve selective mismatching of the transistor to its source and load impedance to accomplish its intended purpose. Only the maximum gain amplifier requires a conjugate matching network design. The specific matching techniques for each category of amplifiers are listed as:

7.2 Maximum Gain Amplifier Design

This section covers the design of the maximum gain amplifier at 2.35 GHz using the SHF-0189 transistor. The SHF-0189 is a high performance Hetrostructure FET (HFET) housed in a surface-mount plastic package (SOT-89) as shown in Figure 7-1. The maximum gain amplifier is conjugate matched at the input and output resulting in very good return loss over a narrow bandwidth. A device can only be conjugate matched if it is unconditionally stable [1]. If the device is potentially unstable in the band of interest, a conjugate impedance match cannot be realized unless we add additional circuitry to stabilize the device.

Figure 7-1 GaAs HFET specifications (courtesy of RF Micro Devices)

7.3 Stabilizing the Device in Genesys

Example 7-1: Measure and display the stability factor, K, the stability measure, B1, and the maximum gain, Gmax, of the SHF-0189 FET transistor from 100 to 8000 MHz. Design the stability network if the transistor is unstable.

Solution: An amplifier design begins with the placement of the device S parameters in a schematic design workspace as shown in Figure 7-2.

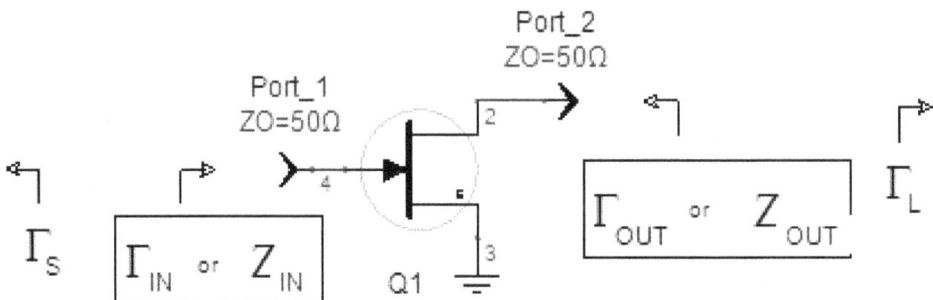

Figure 7-2 Transistor input and output reflection coefficient

The Genesys software has built in functions to determine the stability factor and stability measure as well as Γ_{MS} and Γ_{ML} and the conjugate match reflection coefficients. Sweep the frequency range of the device over the entire range of frequencies contained in the S parameter file. Plot the parameters K, B1, and GMAX for the SHF-0189 device.

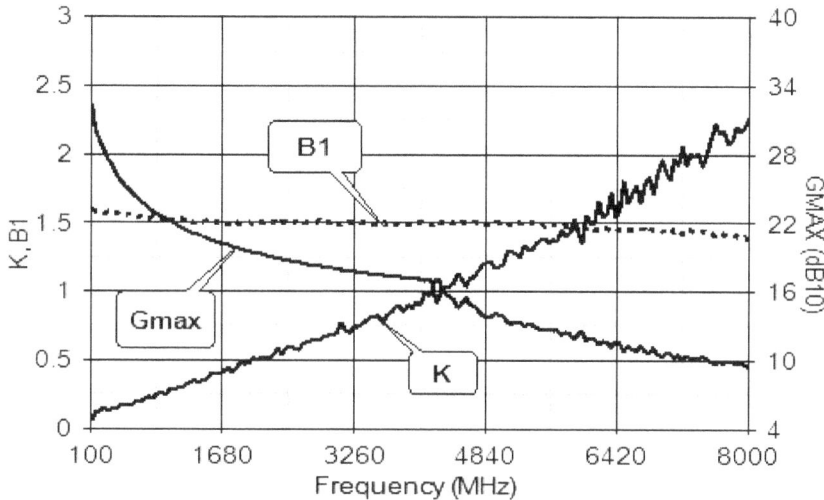

Figure 7-3 Plot of stability parameters and GMAX for the SHF-0189 device

Note that stability factor k is less than 1 from 100 MHz up to about 4500 MHz indicating that the transistor is potentially unstable over most of its frequency range including the design frequency of 2350 MHz.

Example 7-2: Design the stability network for the SHF-0189 device from 100 to 8000 MHz.

Solution: An effective stabilization network for medium and high power transistors employs a parallel RC circuit on the input of the device as shown in Figure 7-4.

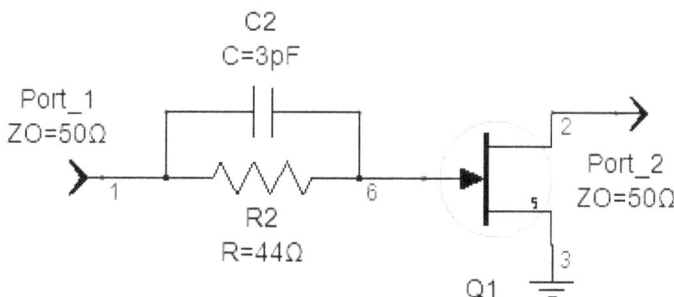

Figure 7-4 Parallel RC stability network

Make both the R and C values tunable with a starting value of 10 Ω and 10 pf respectively. Tune the resistance and capacitor values until K > 1 for frequencies above 500 MHz. The values shown in Figure 7-4 provide sufficient stability above 500 MHz.

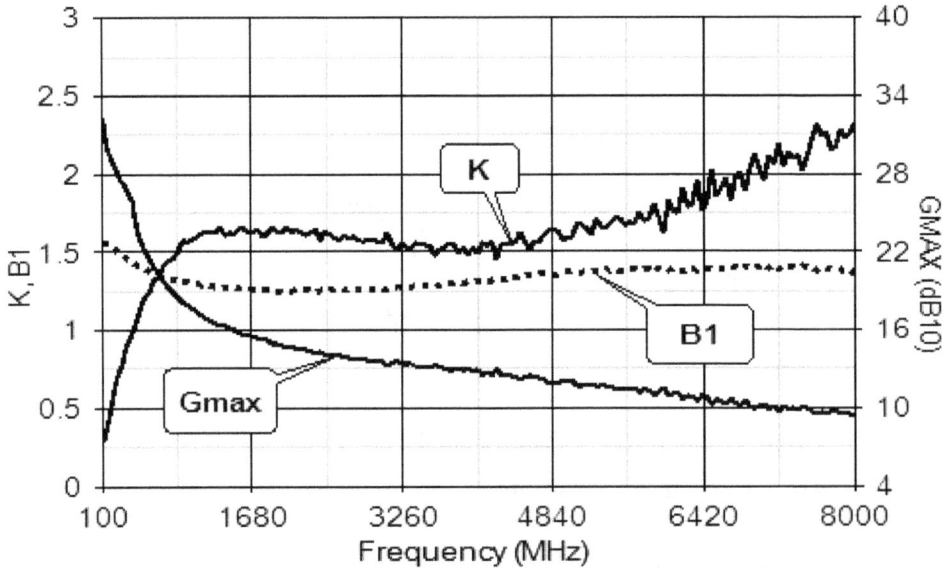

Figure 7-5 Stability parameters and GMAX with parallel RC network

As seen in Figure 7-5, the stability measure, B1, is greater than zero throughout the entire frequency range. However the stability factor, K, goes below one at frequencies from less than 427 MHz. We want to make sure that the device is stable at all frequencies so that there is no possibility of any value of Γ_{ML} or Γ_{MS} that could make the device unstable at these out-of-band frequencies. Therefore add an R-L network in shunt with the input of the transistor for additional low frequency stability. This R-L network can be absorbed into the bias feed for the gate voltage of the transistor. Choose a fixed value of 50 Ω for the resistor. Select a starting value of 100 nH for the inductor and make this value tunable. This R-L network will load the previous R-C network so its values will need to be changed. The inductor's value can be reduced which in turn reduces the low frequency gain while not significantly reducing the in band gain at 2.35 GHz. Figure 7-6 shows the final stabilization network for the device.

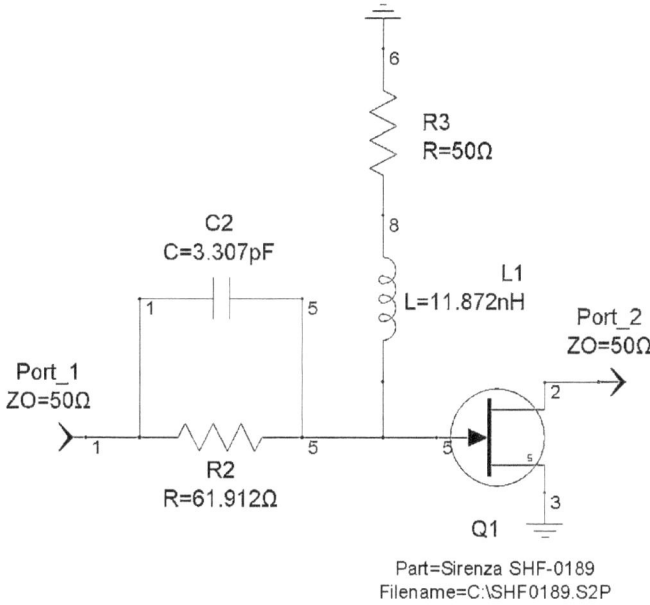

Figure 7-6 Stabilization network for the SHF-0189 transistor

As shown in Figure 7-7, GMAX = 15.975 dB and K = 1.271 at the design frequency of 2.35 GHz. The device is now unconditionally stable for all frequencies from 50 MHz to 8 GHz. Now that the device is unconditionally stable, we can determine the simultaneous conjugate match source and load impedance, Γ_{MS} and Γ_{ML}.

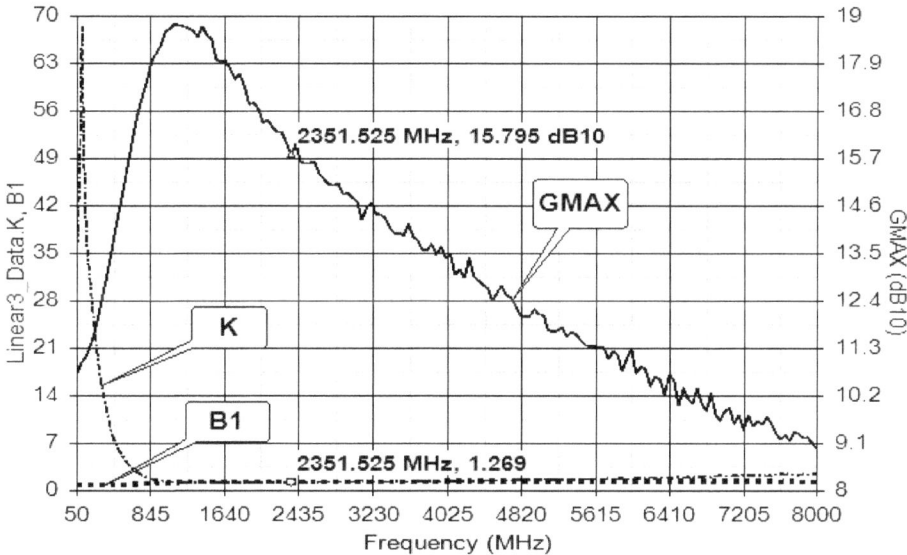

Figure 7-7 Stability parameters and G_{MAX} of stabilized device

7.4 Finding Simultaneous Match Impedances

Genesys has built in functions for the calculation of the Γ_{MS} and Γ_{ML} reflection coefficients as well as the corresponding impedance Z_{MS} and Z_{ML}. Using the schematic of Figure 7-6 create a tabular output of the simultaneous conjugate source and load parameters as shown in Figure 7-8.

Measurement	Label (Optional)	Units	Complex Format	Hide?
GMAX		dB10	(default)	☐
GM1		dB	Mag abs+Angle	☐
GM2		dB	Mag abs+Angle	☐
ZM1			Real+Imaginary	☐
ZM2			Real+Imaginary	☐
			(default)	☐

	F (MHz)	GMAX (dB10)	mag (GM1) (dB)	ang (GM1)	mag (GM2) (dB)	ang (GM2)	re(ZM1) ()	im (ZM1)	re(ZM2) ()	im (ZM2)
1	2350	15.791	0.793	131.771	0.605	141.09	6.918	22.023	13.72	16.471

Figure 7-8 Tabular output functions showing G_{MAX}, Γ_{MS}, Γ_{ML}, Z_{MS}, and Z_{ML}

Note the equivalence between the Genesys function notation and the respective reflection coefficient and impedance.

$$GM1 = \Gamma_{MS} = 0.793 \angle 131.7 \qquad ZM1 = Z_{MS} = 6.92 + j22.02 \ \Omega$$

$$GM2 = \Gamma_{ML} = 0.605 \angle 141.1 \qquad ZM2 = Z_{ML} = 13.72 + j16.47 \ \Omega$$

7.5 Input Matching Network Design

In the design of the SHF-0189 amplifier, we will use the analytical impedance matching techniques developed in chapter 5 and the L-Network Synthesis Utility to generate the matching networks.

When performing the analytical match we are matching the 50 Ω source impedance to the impedance looking into the device impedance (Z_{IN} or Z_{OUT} of Figure 7-7; not the Z_{MS} or Z_{ML}. Z1 = Z_{IN} and Z2 = Z_{OUT} is the impedance looking into the input and output of the device. Z_{MS} and Z_{ML} are the source and load impedances that the device is looking into.

Therefore the complex impedances that we are matching are given as the conjugate of Z_{MS} and Z_{ML}.

$$Z1 = Z1^* sm = 6.92 - j22.02 \ \Omega$$
$$Z2 = Z2^* sm = 13.72 - j16.47 \ \Omega$$

To match the complex load impedance to a resistive source first calculate r and x by normalizing the load impedance with respect to the source resistor, and then determine the matching networks on the basis of the conditions summarized in Table 5-1. The conditions are repeated here.

- If $r < 1$ and $r^2 + x^2 - r < 0$ there exists only two matching networks obtained from Equations (5-21) through (5-24) in the Appendix B.

- If $r < 1$ and $r^2 + x^2 - r > 0$ there exists four matching networks obtainable from Equations (5-21) through (5-24) and Equations (5-26) through (5-29) in the Appendix B.

- If $r > 1$ there exists only two matching networks obtained from Equations (5-26) through (5-29) in the Appendix B.

Example 7-3: Design the maximum gain amplifier input matching network.

Solution: For the input matching network $r = 0.1384$ and $x = -0.44$, therefore, $r < 1$ and $r^2 + x^2 - r = 0.074 > 0$. According to the above conditions, there exists four L-networks that can be used for the input matching network. To calculate the element values of the first matching network we can utilize the Equation Editor in Genesys and follow the procedure:

1. Enter the source and load impedance and design frequency

2. Normalize the load impedance

3. Write the appropriate equations to calculate B and X

4. Based on the positive or negative values of B and X, calculate the element values of the matching network as in Figure 7-9.

```
Z0=50
RL=6.918
XL=-22.03
f=2.35e9
r=RL/Z0
x=XL/Z0
B2=-sqrt((1-r)/r)/Z0
X2=-Z0*(sqrt(r*(1-r))+x)
L1=X2/(2*pi*f)
L2=-1/(2*pi*f*B2)
```

Figure 7-9 Equation Editor calculating matching element values

To display the response of the input matching network, add a Schematic in Genesys and place the matching elements with the values in the Figure 7-10. Add a Linear Analysis from 1350 MHz to 3350 MHz to analyze the response of the matching network, S11, and S21 both in dB. The swept insertion loss and return loss are also shown in Figure 7-10.

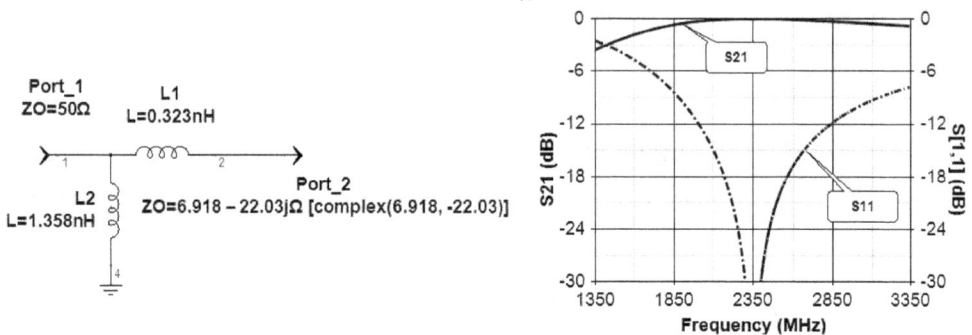

Figure 7-10 Impedance matching network for the input match

7.6 Output Matching Network Design

Example 7-4: Design the output matching network for the SHF-0189 at 2350 MHz.

Solution: Enter Z2 for the complex impedance at Port 2 in the L Network Impedance Matching Synthesis Tool. Scrolling through the four possible solutions we find that only two solutions give a good return loss and corresponding impedance match for the output network. The solution of Figure 7-11 is selected for the output matching network because of the broader band transmission characteristic.

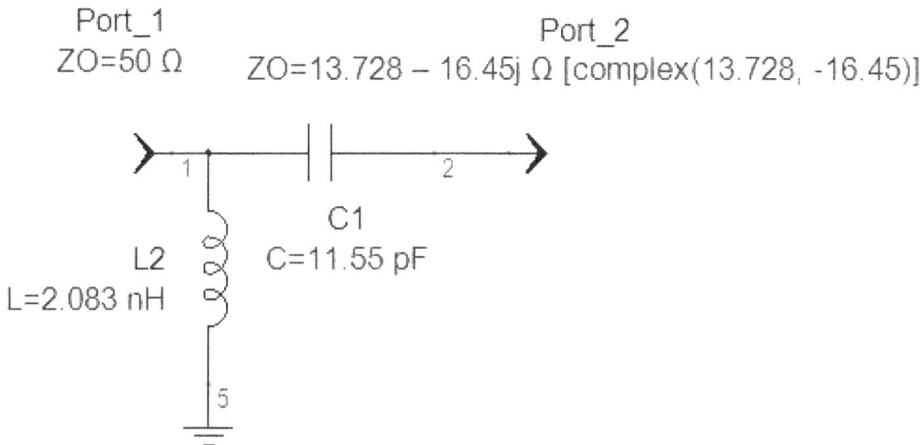

Port_1   Port_2
ZO=50 Ω   ZO=13.728 – 16.45j Ω [complex(13.728, -16.45)]

C1
C=11.55 pF

L2
L=2.083 nH

Figure 7-11 Output matching L-network

7.7 Ideal Model of the Maximum Gain Amplifier

Example 7-5: Assemble and simulate the ideal maximum gain amplifier. Use a linear sweep from 100 MHz to 5000 MHz to analyze the response of the amplifier.

Solution: Create the ideal model of the amplifier by attaching the input and output matching networks to the stabilized device, as shown in Figure 7-12. Use a Linear Sweep from 100 MHz to 5000 MHz to analyze the response of the amplifier. The swept gain, input return loss, and output return loss is shown in Figure 7-13. As the return loss ishows, an excellent match has been achieved at the input and output ports. Also the maximum gain (15.79 dB) of the amplifier at 2.35 GHz is exactly the originally calculated G_{MAX} indicating a successful simultaneous conjugate impedance match.

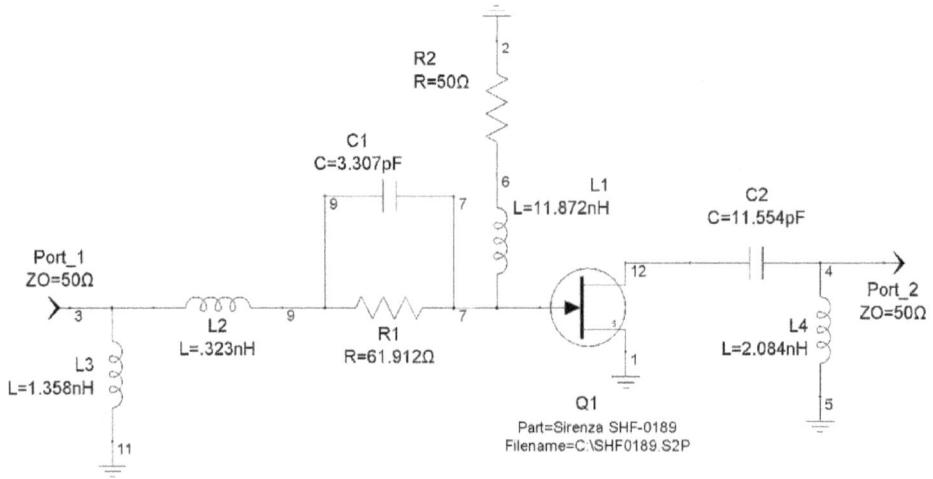

Figure 7-12 Schematic of the ideal amplifier circuit

The simulated response is show in 7-13.

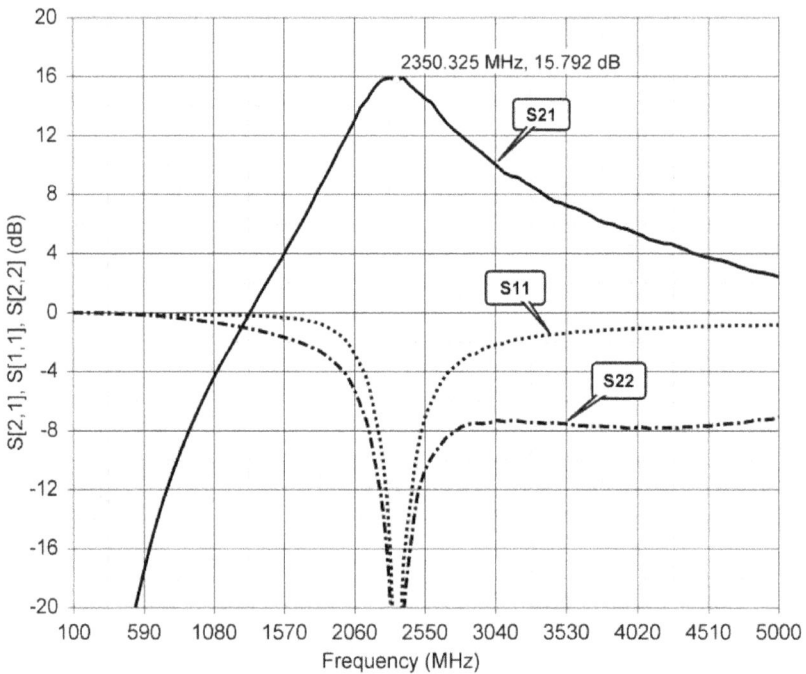

Figure 7-13 Swept response of the ideal amplifier design

The fractional bandwidth of the amplifier is calculated by sweeping the response over a narrower frequency range as shown in Figure 7-14. Markers.

are placed on the passband response at the 0.1 dB ripple bandwidth. As Figure 7-14 shows the 0.1 dB bandwidth is:

$$BW_{20dB} = 2396\text{-}2333 = 63 \text{ MHz}.$$

This corresponds to a fractional bandwidth of about 2.66 % consistent with a narrowband amplifier. The conjugate match is achieved at the single frequency of 2.35 GHz.

$$FBW_{20dB} = \frac{63}{\sqrt{(2396).(2333)}} = 2.66 \%$$

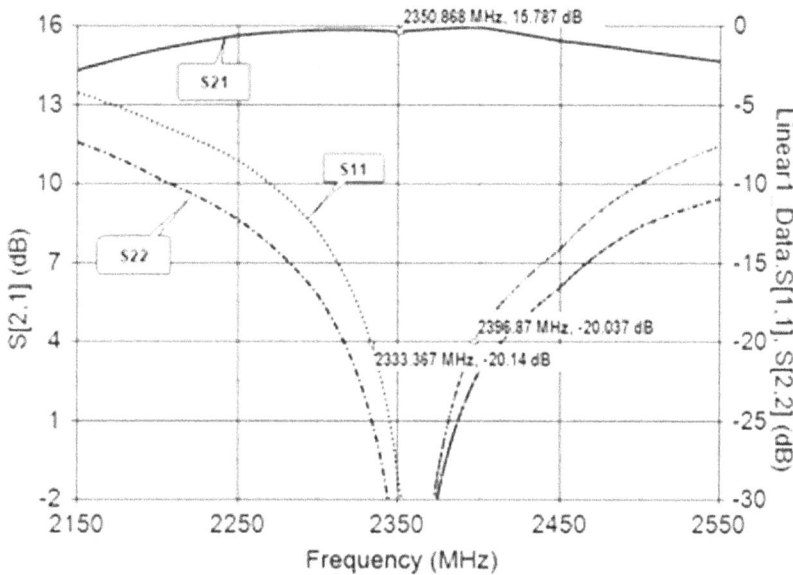

Figure 7-14 Ideal amplifier response with markers at 20 dB return loss

7.8 Physical Amplifier Design and Layout

Example 7-6: Design the physical amplifier bias feed and the physical PCB layout for the SHF-0189 maximum gain amplifier.

Solution: Start the physical design of the amplifier by placing the device S parameter file on the schematic. Use a microstrip Via Hole model for the connection of the transistor source lead to ground. Specify a via hole radius of 12 mils. Then assign the artwork replacement element for the SHF-0189 by opening the Properties window of the S parameter file. Select the Change Footprint button and browse to the location of the artwork replacement element. Proceed with the construction of the stabilization network portion of the design. Use Modelithics surface mount components (30 mil x 60 mil) for the stabilization components as these will be easy to tune or optimize Use the S parameter files of the components that we do not intend to optimize. A 470 pF capacitor is added as a gate bypass capacitance. On the drain side of the transistor a 33 nH inductor is used as an RF choke to supply the drain voltage. Edit the footprint of the S parameter files and change the footprint to a 0603 surface mount component from the Genesys library. The Modelithic components will automatically have the correct footprint assigned to its properties.

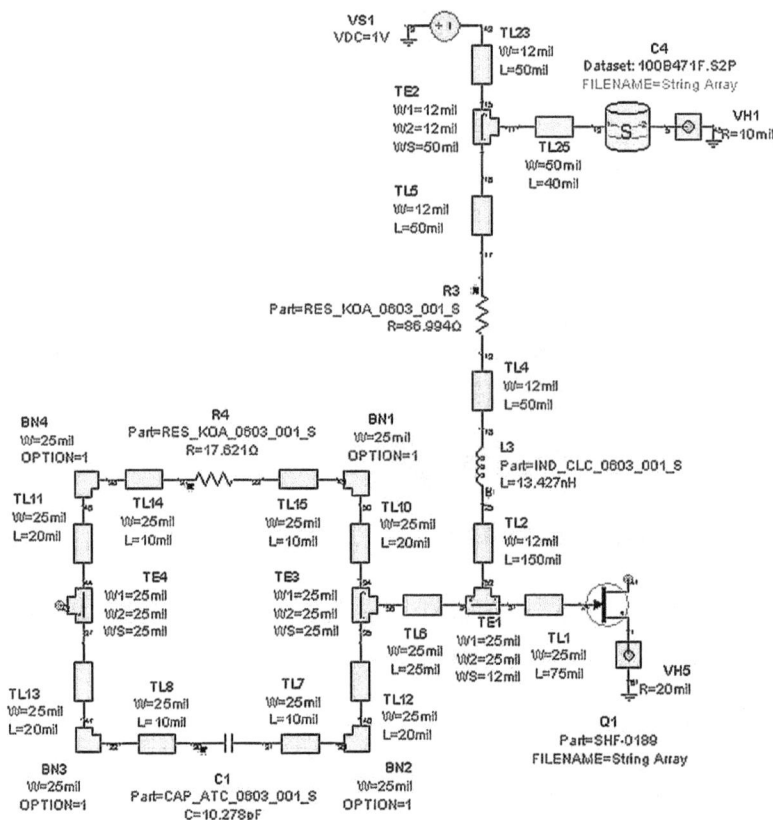

Figure 7-15 Construction of physical amplifier model

Continue with the physical model construction by adding the matching networks. Note that on the input side of the circuit the matching network has an inductor connected to ground. Because there will be a negative voltage applied to the gate of the transistor, a DC blocking capacitor must be added to keep the inductor from shorting out the gate bias. Add an S parameter file for a 1.5 pF chip capacitor to act as a DC block. The capacitance value should be chosen such that the capacitor is operating in series resonance for minimum insertion loss. Knowing that the chip capacitor typically has some series inductance, the small series inductance of 0.323 nH can be absorbed into the DC blocking capacitor. The output matching circuit includes a series capacitance so no additional DC blocking capacitor in needed. The completed schematic of the physical amplifier model is shown in Figure 7-16. The physical PCB layout is shown in Figure 7-17.

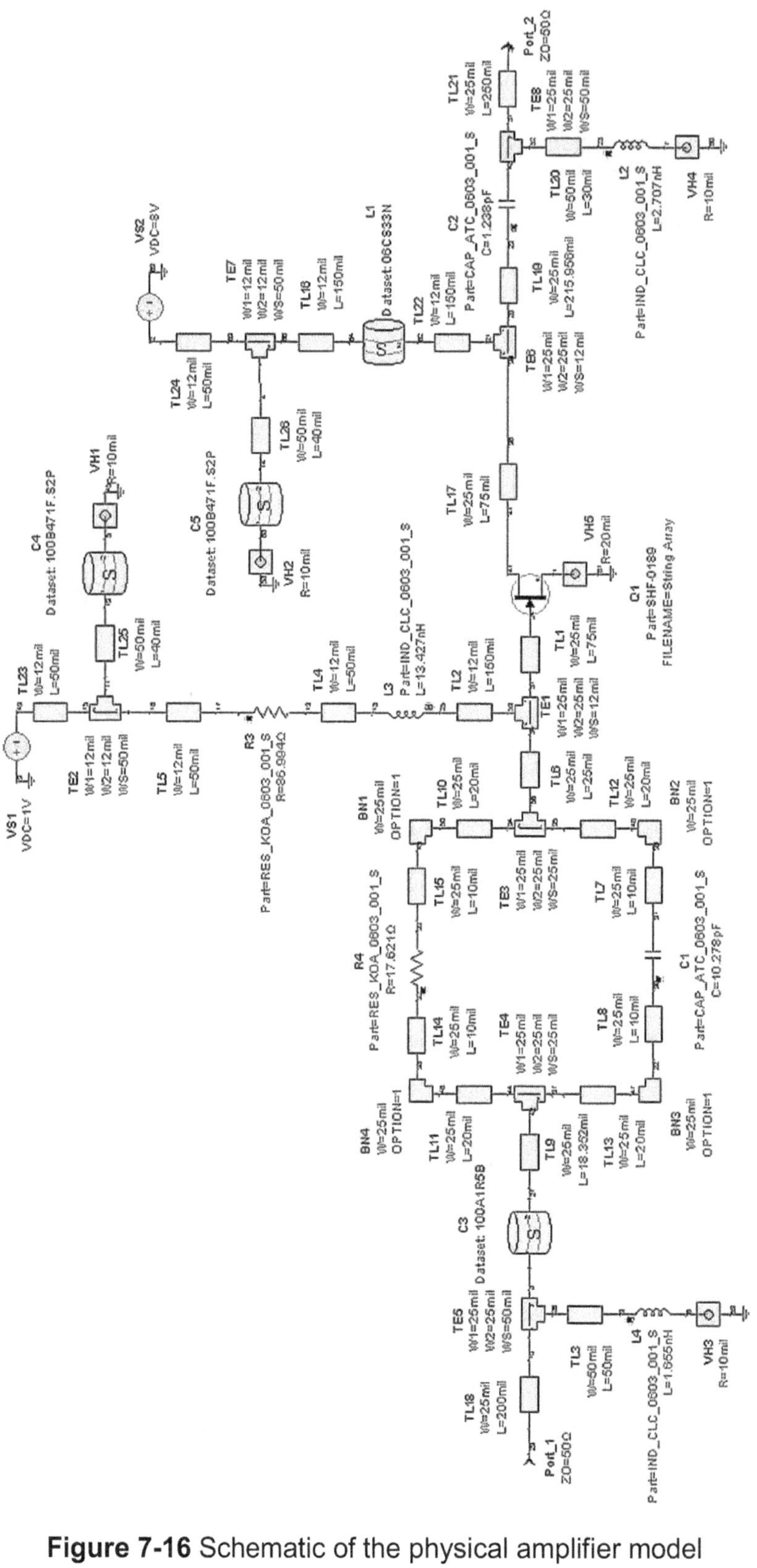

Figure 7-16 Schematic of the physical amplifier model

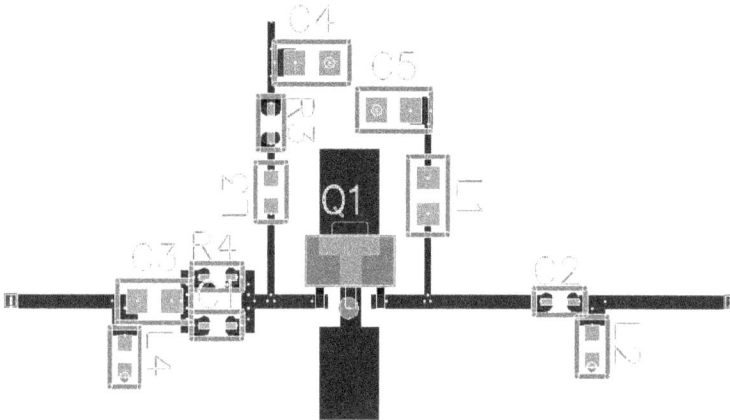

Figure 7-17 Physical layout of completed amplifier circuit

7.9 Optimization of the Amplifier Response

After completion of the physical circuit schematic of Figure 7-22 the circuit response is swept from 100 MHz to 5000 MHz, as shown in Figure 7-24. The response of the amplifier has significantly changed from the response of the ideal amplifier response of Figure 7-15. The gain and return loss have shifted in frequency. This is a typical result of replacing ideal elements with real physical models. This ability to deal with package parasitics and physical layout is a very powerful benefit of modeling the amplifier in Genesys. We know that it is possible to make the component element values tunable and manually tune each element while observing the change in response. This is quite useful and encouraged so that you can see the circuit sensitivity to particular components. As the number of components and tunable elements increase in a given design, the tuning of individual elements can be cumbersome. In this section we will introduce the use of Optimization to tune the amplifier's response to the desired characteristics. We can think of Optimization as an automatic circuit tuner. Selected components in the amplifier can be assigned as tunable variables. Multiple goals can then be set such as gain levels, return loss, etc. The optimization is then run and the Genesys software will execute a mathematical optimization algorithm to determine the values of the variables required to achieve the specified goals. In this example set all seven of the Modelithics model values as tunable. Also make the microstrip line lengths for the series sections TL9 and TL19 variable. We will let the software optimize these variables to achieve the desired response. Make sure to set response goals for out-of-band responses as well as in band response.

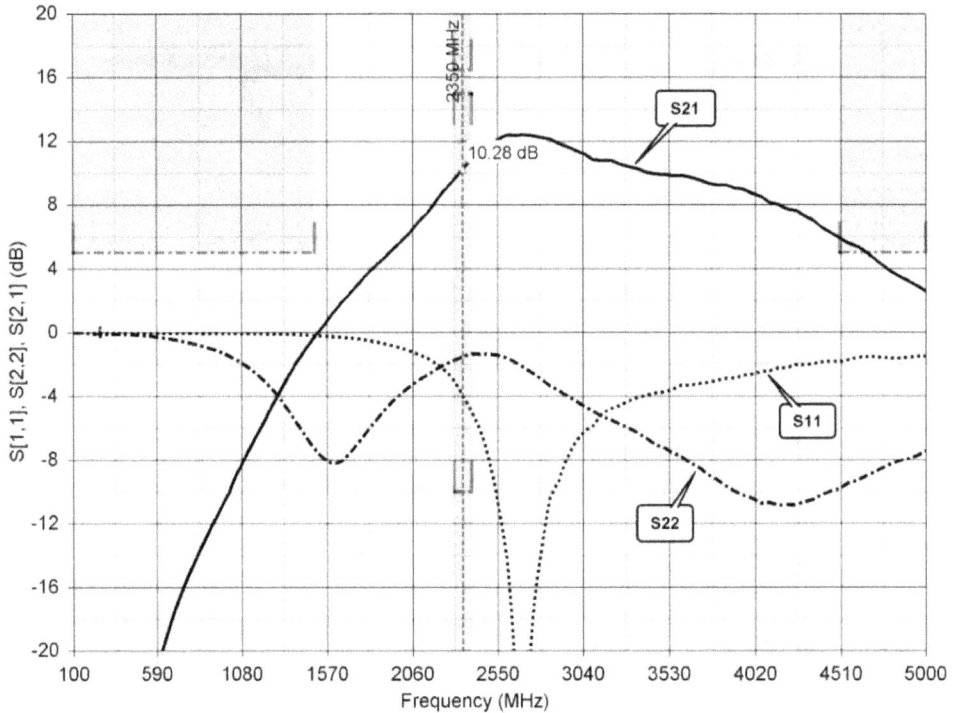

Figure 7-18 Initial response of the physical amplifier circuit of Figure 7-22

7.10 Specific Gain Amplifier Design

In the design of RF and microwave amplifiers, it is common to design with potentially unstable transistors. The designer must know how to deal with potentially unstable devices. This section outlines the design of a 2.3 GHz amplifier using the RT243 GaN HEMT device. The RT243 has a usable frequency range of 100 MHz to 5 GHz.

Example 7-7: Analyze the stability condition for the RT243 device.

Solution: Examine the stability circles for the RT243 by creating a new workspace and a new schematic in Genesys with the small signal S parameter file as shown in Figure 7-19. Setup an S parameter simulation with both the start and stop frequencies set at 2.3 GHz.

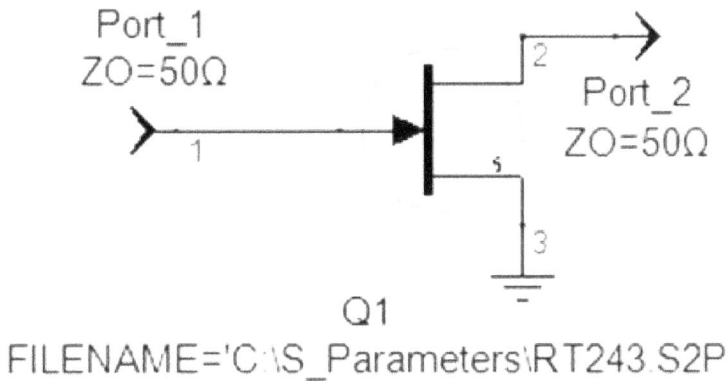

Port_1
ZO=50Ω

Port_2
ZO=50Ω

Q1
FILENAME='C:\S_Parameters\RT243.S2P

Figure 7-19 Schematic with RT243 S parameter file

A sweep of the circuit will result in a plot of the center of each stability circle displayed. To display the entire circle, click on the center trace (dot) on the Smith Chart. As Figure 7-20 shows the circles are far away from the center of the chart therefore making the outside of the circles the stable region. There is only a small region of the output stability circle that represents an unstable region of impedance. These unstable regions are shaded as shown in Figure 7-20. If the device were unconditionally stable the stability circles would lie completely outside the circumference of the Smith Chart. Add another Linear Analysis from 100 MHz to 3000 MHz with 100 MHz steps. Then add a Table to display K, B1, and GMAX. Note that stability factor k is less than 1 indicating that the transistor is potentially unstable over its entire frequency range. G_{max} for the device without a stabilization network is 22.6 dB at 2.3 GHz. We want to design the amplifier to have a gain of 16 dB with a 50 Ω source and load impedance.

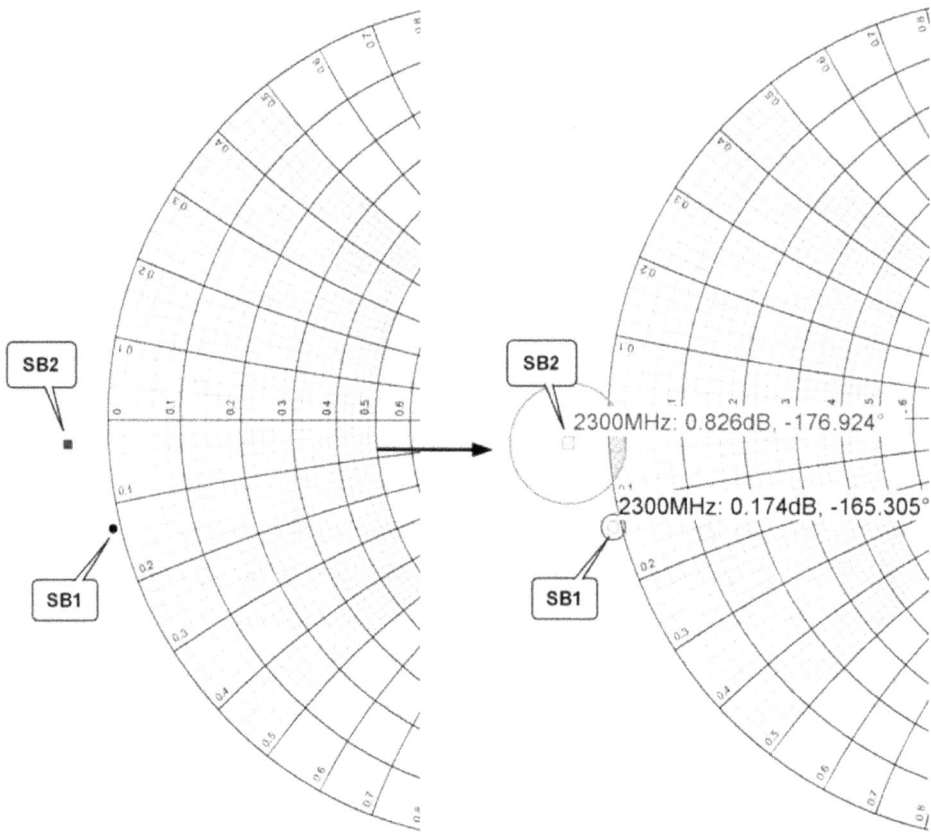

Figure 7-20 Stability circle centers and full circles at 2.3 GHz

The stability parameters are shown in Table 7-2.

F (MHz)	K ( )	B1 ( )	GMAX (dB10)
100	0.272	0.966	35.753
300	0.122	0.838	30.862
500	-0.236	0.904	28.363
700	0.266	0.853	27.462
900	0.316	0.845	26.4
1100	0.352	0.826	25.528
1300	0.379	0.8	24.778
1500	0.419	0.769	24.125
1700	0.43	0.739	23.546
1900	0.463	0.711	23.035
2100	0.537	0.69	22.75
2300	0.654	0.666	22.61
2500	0.802	0.641	22.448
2700	0.861	0.625	22.188
2900	0.931	0.609	21.911
3000	0.97	0.601	21.765

Table 7-2 RT243 device stability factors and GMAX

Fortunately Genesys has the ability to plot a series of constant gain circles. This enables the plot of the circles on the Smith Chart without the need to solve the Equations (7-5) and (7-6). Selecting the GP function (power gain circles) in the output graph plots a series of seven gain circles. When $K < 1$ the smallest circle represents the maximum gain, GMAX, and the inside of the circle is shaded [3]. This means that impedances inside this circle can be unstable. The successive Power Gain circles represent gain values of GMAX -1dB, GMAX-2dB, GMAX-3 dB, GMAX-4 dB, GMAX-5 dB, and GMAX-6 dB. Thus the circles represent the gain values of: 22.6 dB, 21.6 dB, 20.6 dB, 19.6 dB, 18.6 dB, 17.6 dB and 16.6 dB respectively. Figure 7-21 shows that the last or largest circle is the 16.6 dB power gain circle. An examination of the power gain and stability circles reveals another reason why we may not want to use the maximum gain from the device. The highest gain circle is very close to the region of instability on the output stability circle, SB2.

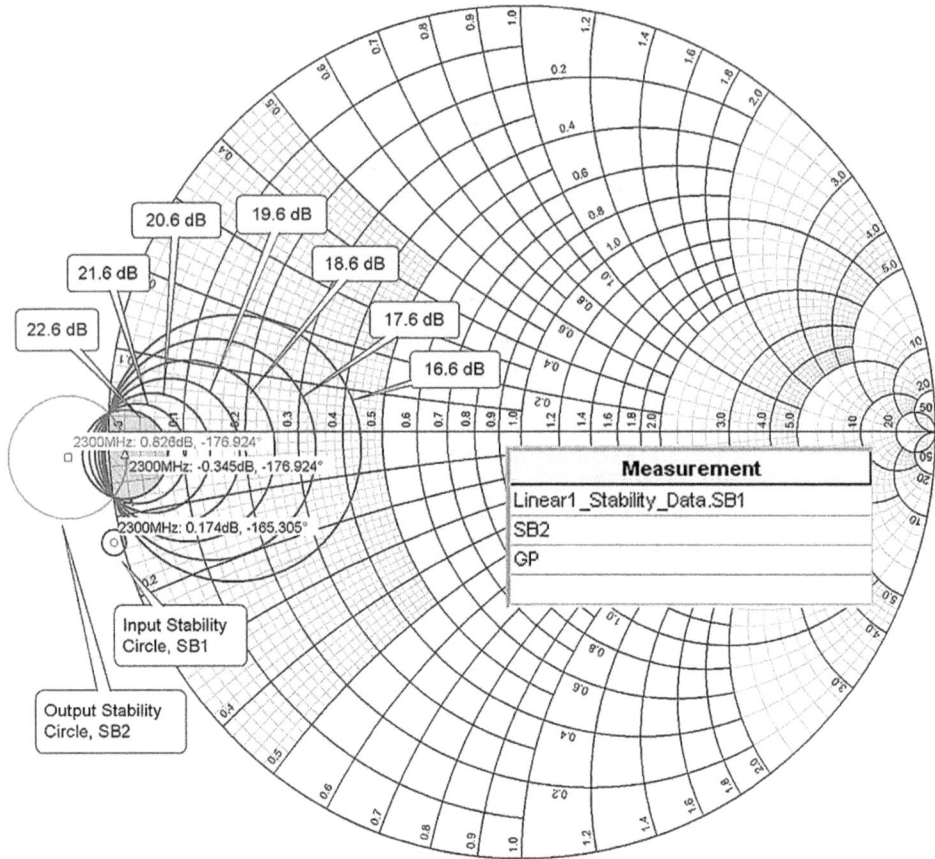

Figure 7-21 Constant power gain circles for the RT243 at 2.3 GHz

We'll choose the 16.6 dB circle because it is the closest to the design goal of 16 dB. The load impedance can be any point on the circumference of the 16.6 dB circle that does not enter the shaded, unstable, region. It is a good practice to choose locations that are not close to the unstable regions of the stability circles. The load impedance can be visually read from the Smith Chart. The load impedance selected on the 16.6 dB Power Gain Circle is shown in Figure 7-22. The impedance at this point is 22 + j0 Ω. To precisely identify this impedance a new design schematic can then be created with an IMP element to define this impedance on the Smith Chart. The IMP element impedance can be plotted on the same Smith Chart as the Power Gain circles. The IMP element's resistance and reactance can be tuned to get the impedance to exactly overlay on the desired point on the circumference of the 16.6 dB Power Gain circle.

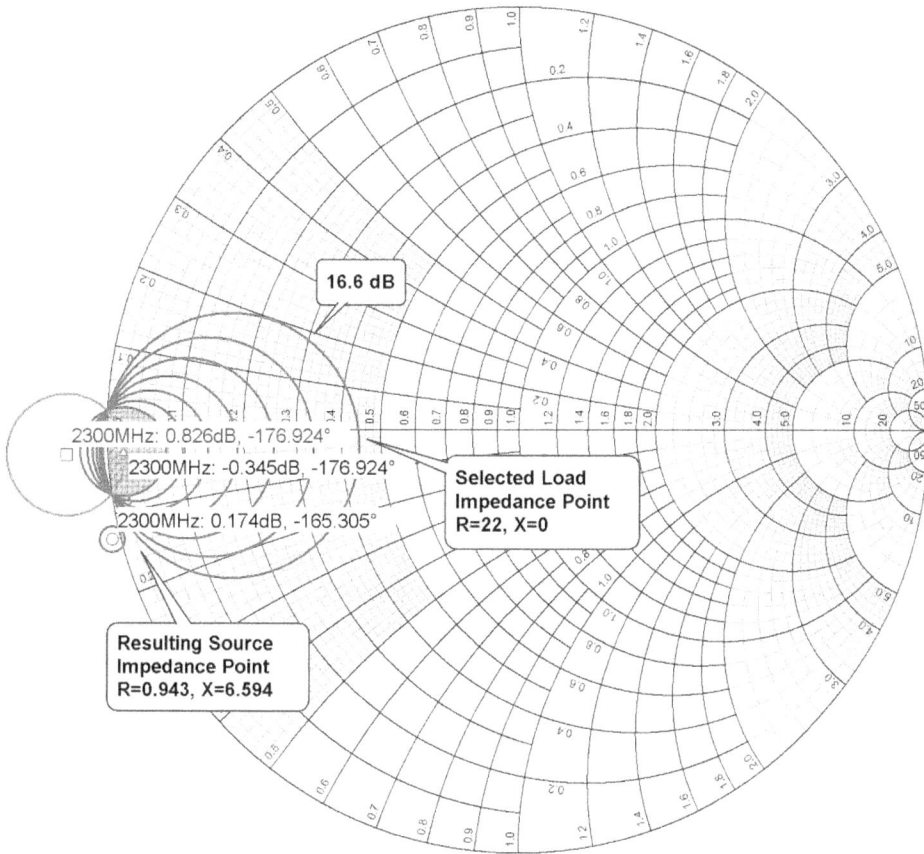

Figure 7-22 Location of source and load impedance for power gain match

Once the load impedance has been selected it is necessary to calculate the resulting source reflection coefficient that will achieve a conjugate match on the input. The following equation is used to calculate this source reflection coefficient. The source reflection coefficient is therefore defined as a function of the chosen load reflection coefficient, Γ_L. Note that the input match is then a conjugate match of the calculated source reflection coefficient, Γ_S.

$$\Gamma_S = \left[S_{11} + \frac{S_{12}S_{21}\Gamma_L}{1 - \left(\Gamma_L S_{22}\right)} \right]^*$$

The chosen load impedance must first be converted to a reflection coefficient, Γ_L. This is easily handled by the Equation Editor, as shown in Figure 7-23. Simply define the Load variable as the reflection coefficient S[1,1]. Once the load impedance is defined as a polar formatted variable, Equation (7-18) can be solved within the same Equation Editor. Because two frequency step points were needed to display the circles, the equations are entered as swept array variables. The source impedance is calculated as 0.943-j6.594 as shown in Figure 7-23. Location of the source and load impedance for power gain match is shown in Figure 7-22.

```
RS=50
RL=318.6
XL=-166.1
f=432e6
x=XL/RS
t1=(x+sqrt(r*(r^2+x^2-2*r+1)))/(r-1)
t2=(x-sqrt(r*(r^2+x^2-2*r+1)))/(r-1)
d1=360*(pi+atan(t1))/(2*pi)
d2=360*(pi+atan(t2))/(2*pi)
B1=(x*t1^2+(r^2+x^2-1)*t1-x)/(RS*(r^2+x^2+t1^2+2*x*t1))
B2=(x*t2^2+(r^2+x^2-1)*t2-x)/(RS*(r^2+x^2+t2^2+2*x*t2))
so1=360*(pi-atan(RS*B1))/(2*pi)
so2=-360*atan(RS*B2)/(2*pi)
```

Figure 7-23 Solution of Equation (7-18) in Equation Editor

7.11 Design of the Impedance Matching Networks

The simple two-element LC matching circuit can be synthesized for both the input and output matching circuits.

Example 7-8: Design the output matching network for the RT243 transistor at 2300 MHz.

Solution: The simple two-element LC matching circuit can be synthesized for both the input and output matching circuits. Setup a design schematic for the design of the output matching circuit. Use the generic impedance element, IMP, to define the load impedance. Then place a 50 Ω resistor to represent the 50 Ω output impedance in which the load impedance is to be matched. Place a capacitor, C1, in shunt with the 50 Ω resistor. Initially set the inductor value, L1, to 0 nH. Then tune the value of the capacitor, C1, until the resulting impedance is positioned on a constant resistance circle that is aligned with the load impedance. Then tune the inductance value to bring the resulting impedance to overlay the load impedance. The resulting LC values are shown in Figure 7-24.

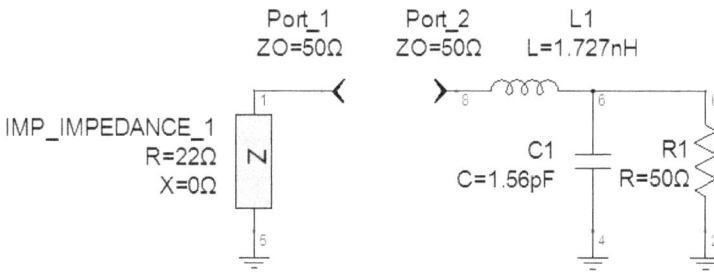

Figure 7-24 Schematic used to design the output matching circuit

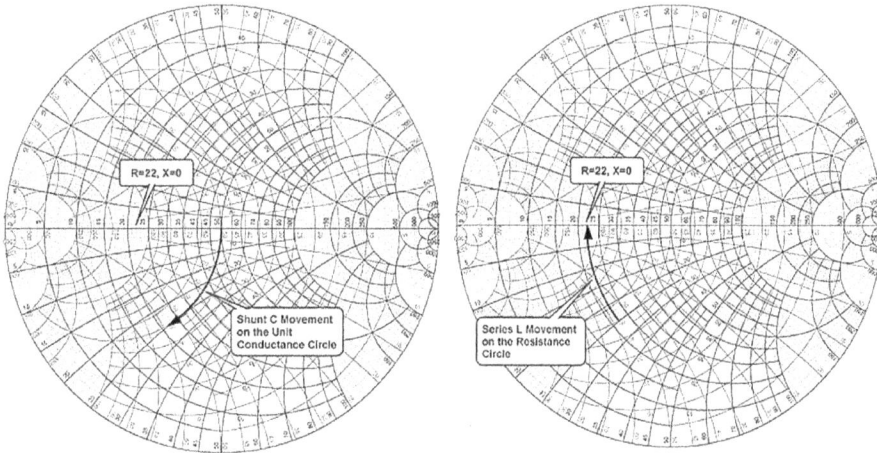

Figure 7-25 Shunt C and series L, output impedance match progression

Example 7-9: Design the input matching network for RT243 at 2300 GHz.

Solution: In similar fashion the input matching circuit is designed by initially setting the inductor value of L1 to zero. Tune the capacitance value to a constant resistance circle near the source impedance. In this case the circuit is almost matched. Only a very small inductance, 0.014 nH, is required to complete the match. Here we see that the source impedance is located very near to the input stability circle. Very careful component selection would be required to make sure that the matching circuit would not push the match into the area of unstable performance. Any component tolerance or drift over temperature may result in an unstable amplifier circuit. It is this reason why an RC stability circuit is sometimes preferred.

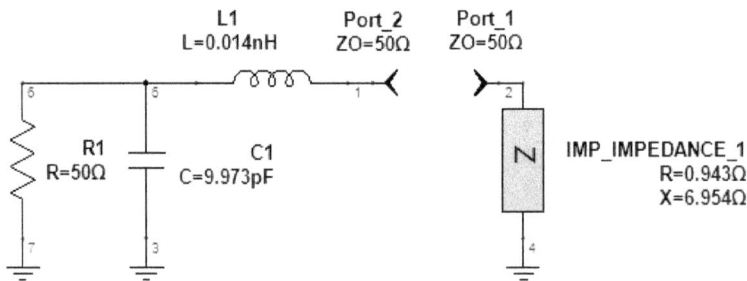

Figure 7-26 Schematic used to design the input matching circuit

The shunt C–series L, input impedance match progression is shown in Figure 7-27.

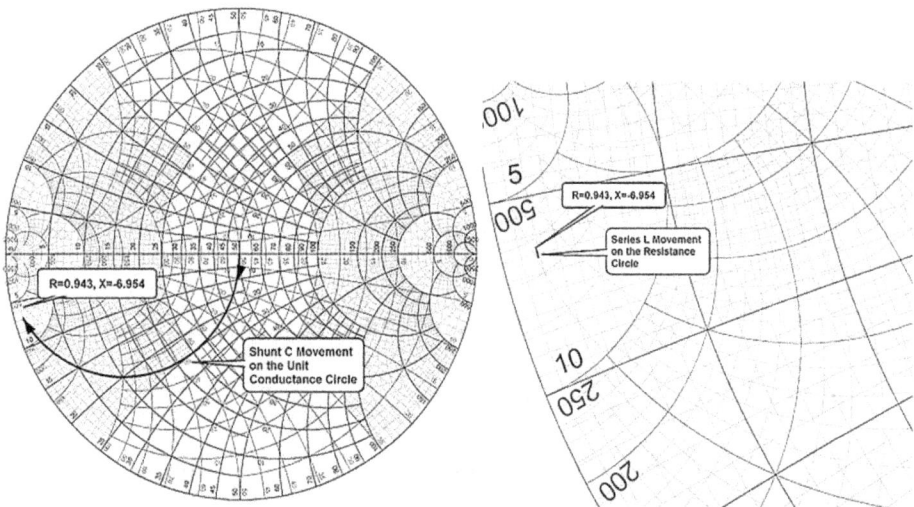

Figure 7-27 Shunt C–series L, input impedance match progression

7.12 Assembly and Simulation of the Amplifier

Example 7-10: Assembly and Simulation of the Specific Gain Amplifier.

Solution: Create a new schematic and attach the input and output matching circuits to the device S parameter file. Create a Linear Analysis and sweep the amplifier from 1.3 GHz to 3.3 GHz. The resulting response is shown in Figure 7-29. Note that the gain at 2300 MHz is 16.64 dB verifying that the Power Gain circle match has been successful. Also note a major difference in return loss compared to the simultaneous conjugate match case. Here we see that the output return loss is rather poor, -1.39 dB. This is a result of the selective mismatching process to achieve a much lower value of gain than the GMAX of the device. The input return loss is very good because a conjugate match of the resulting device and load impedance combination was performed.

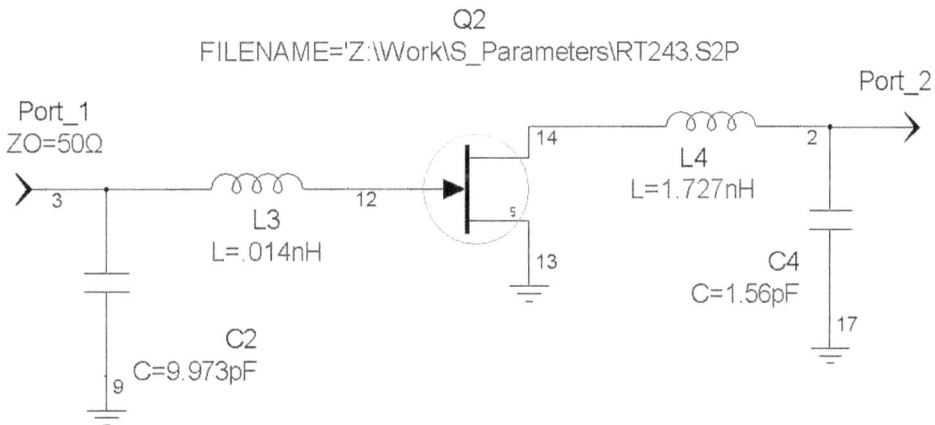

Figure 7-28 Ideal amplifier schematic for power gain matching example

Begin with the package outline drawing of the transistor as was done in Section 7.3. Then progress with the microstrip PCB tracks and physical models for the inductors and capacitors. Clearly the 0.014 nH series inductor and possibly the 1.727 nH inductor will be absorbed into the physical design of the circuit. Finally optimize the amplifier to meet the original design goals.

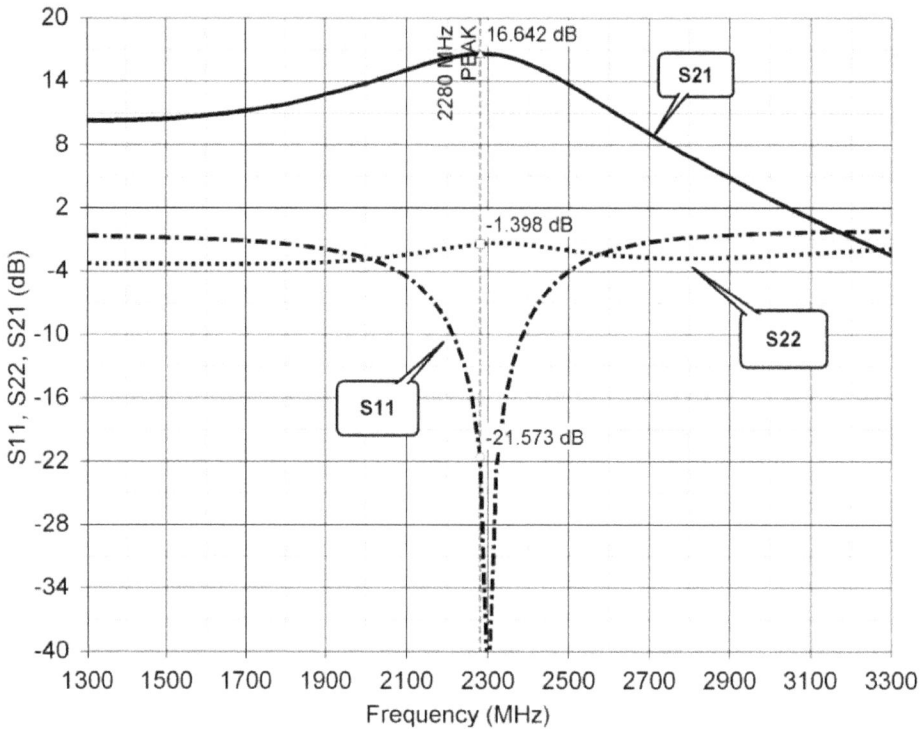

Figure 7-29 Response of the power gain matched amplifier

7.13 Low Noise Amplifier Design

One important case of selectively mismatched amplifier design is the Low Noise Amplifier. In LNA design the input is not matched to the reflection coefficient that results in maximum gain amplifier but rather to a reflection coefficient that gives the desired noise figure. Low noise amplifiers are frequently used as the input stage in a radio receiver or satellite down converter to minimize the noise that is added to the amplified signal.

Center Frequency:	432 MHz
Gain:	20 dB minimum
Noise Figure:	1.2 dB maximum
Output Return Loss:	Less than -10 dB

Calculate the stability and noise parameters for the AT 30511 device

The Avago AT30511 low noise transistor will be used for this design. The S parameter and noise parameter file for this device is shown in Figure 7-30. Create a Genesys schematic with the device S parameter file as shown in Fig 7-30. Add a Table to display the stability parameters K and B1 and the noise parameters Γ_{opt} and Z_{OPT}. Also add a Smith Chart with Γ_{opt} and the stability circles. Then create a Smith Chart and display the stability circles along with the noise circles.

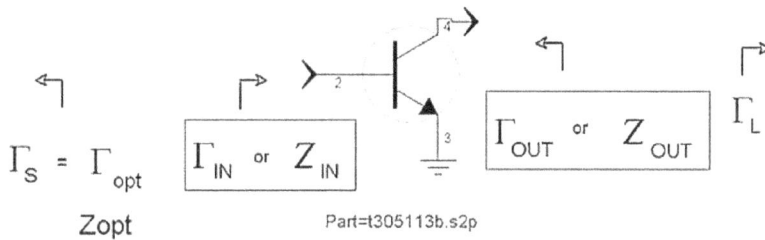

F (MHz)	K ()	B1 ()	mag(GOPT) (dB)	ang (GOPT)	re(ZOPT) ()	im (ZOPT)	NFmin (dB)	GMAX (dB10)
432	0.387	0.274	0.781	7.481	318.445	166.114	1.083	24.974
433	0.387	0.275	0.781	7.503	317.883	166.168	1.083	24.962

Figure 7-30 AT30511 S parameter, K, B1, Γ_{opt}, Z_{opt}, NFmin, and GMAX

Figure 7-30 shows that K < 1 meaning that the device is potentially unstable. This means that there are source and load reflection coefficients that can result in instability and oscillation. Figure 7-31 shows a plot of the input stability circle along with the noise circles and Γ_{opt}. In this example, the inside of the input stability circle, SB1, represents the region of stable source reflection coefficients. Genesys shades the region of the stability circle that represents unstable reflection coefficients. Because the input reflection coefficient, Γ_{opt}, is well inside of the input stability circle, SB1, and we have no specification for input return loss, we can proceed with matching the input to Γ_{opt}. From the discussion of Specific Gain matching we can deduce that because a specific reflection coefficient is selected at the input of the device, the available gain circles can be used to determine the expected gain from the device when matched to Γ_{opt}. Figure 7-32 shows the available gain circles plotted along with Γ_{opt}. Gmax was defined as 24.9 dB.

The available gain circles are drawn for gains of 0, 1, 2, 3, 4, 5 and 6 dB less than maximum gain. Figure 7-32 shows the available gain circles for the AT30511 at 432 MHz. The 21.9 dB circle intersects Γ_{opt} therefore the expected gain of the device when matched to Γ_{opt} is 21.9 dB.

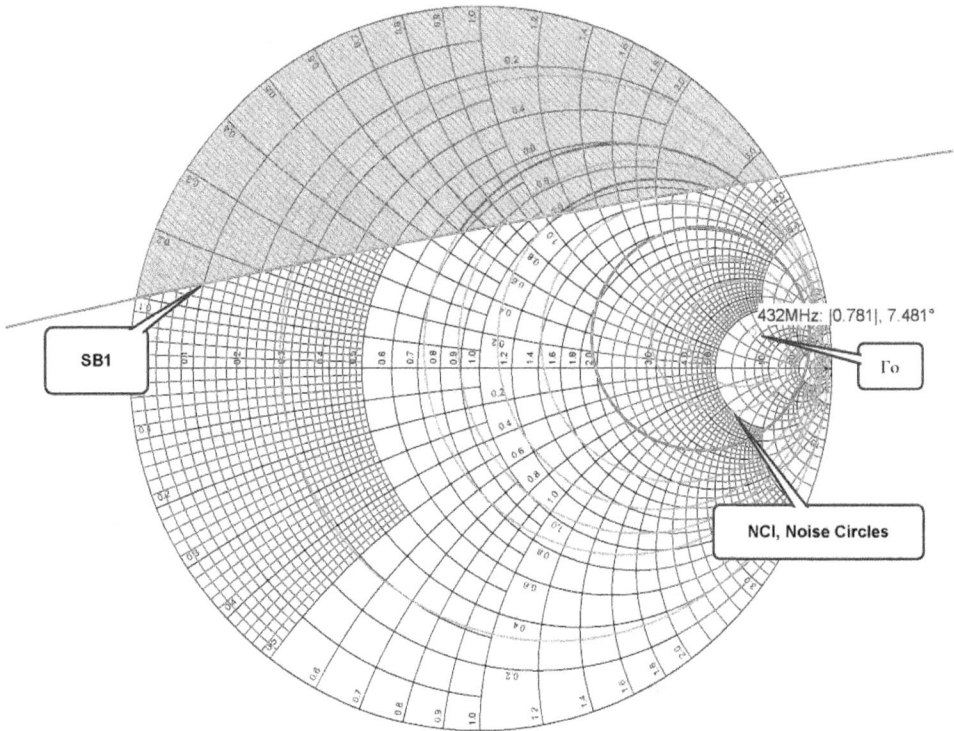

Figure 7-31 Input stability circle with noise circles

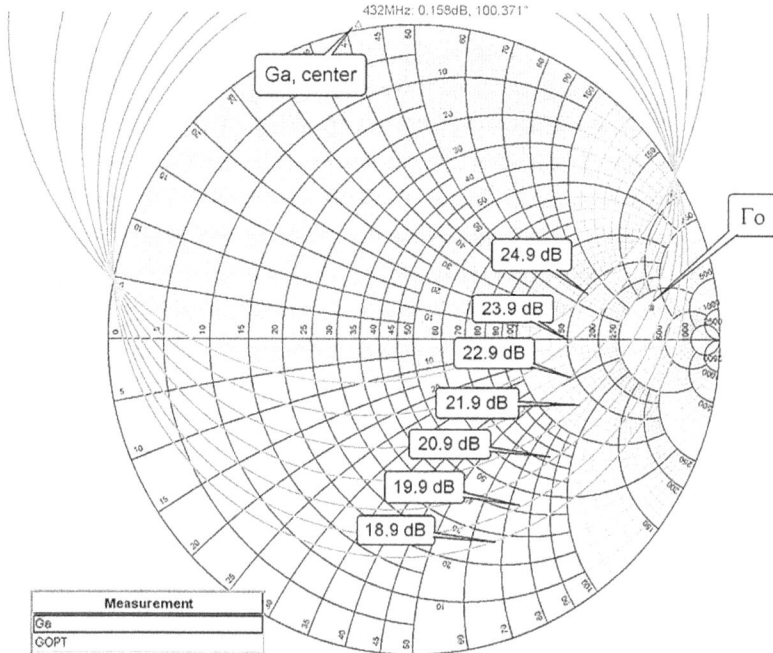

Figure 7-32   Available gain circles and Γ_{opt}

7.14 LNA Input Matching Network Design

The input and output matching circuits will be designed using the single-stub matching networks. Because we are designing essentially a fixed frequency LNA, a simple single-stub matching network can be used.

Example 7-11: Design the LNA Input Matching Network at 432 MHz.

Solution: The input and output matching circuits will be analytically designed using the single stub matching networks. Because we are designing essentially a fixed frequency LNA, a simple single-stub matching network can be used. The equations developed in chapter 6 will be used to determine the input and output matching networks. Referring to Figure 7-43, we match the 50 Ohm source impedance to the input impedance Z_{in} of the device. Z_{opt} is the impedance looking from the device back toward the 50 Ω source. Therefore, in the design of the input matching network, we must use the conjugate of Z_{opt} for the complex load impedance.

Figure 7-33 g ives Z_{opt} = 318.4 + j166.1 ,Ωtherefore, Z_{in} is defined as Z_{opt}^*= 318.4 − j166.1 Ω. As Figure 7-46 shows there are two single-stub matching networks that match the 50 Ωsource impedance to the load impedance.

```
RS=50
RL=318.6
XL=-166.1
f=432e6
x=XL/RS
t1=(x+sqrt(r*(r^2+x^2-2*r+1))))/(r-1)
t2=(x-sqrt(r*(r^2+x^2-2*r+1))))/(r-1)
d1=360*(pi+atan(t1))/(2*pi)
d2=360*(pi+atan(t2))/(2*pi)
B1=(x*t1^2+(r^2+x^2-1)*t1-x)/(RS*(r^2+x^2+t1^2+2*x*t1))
B2=(x*t2^2+(r^2+x^2-1)*t2-x)/(RS*(r^2+x^2+t2^2+2*x*t2))
so1=360*(pi-atan(RS*B1))/(2*pi)
so2=-360*atan(RS*B2)/(2*pi)
```

Figure 7-33 Calculation of the single-stub input matching network

For the input matching network, we select the shorter open-stub length so2 = 68.216 degrees and the corresponding line d2 = 105.58 degrees, as shown in Figure 7-34.

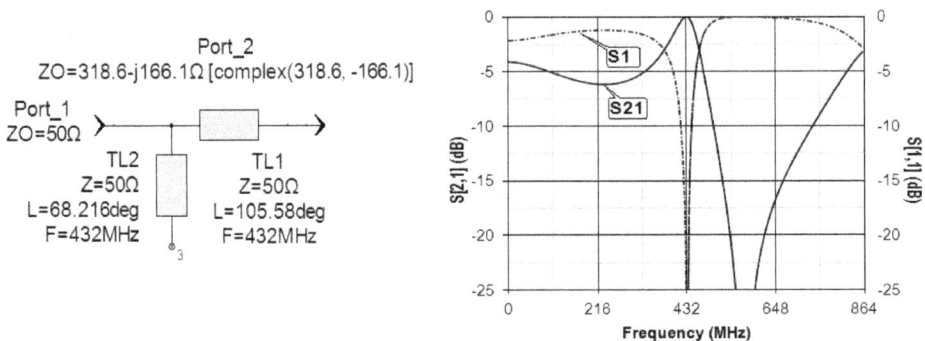

Figure 7-34 Single-stub input matching network and simulated response

7.15 LNA Output Matching Network Design

The output load reflection coefficient, Γ_L, is defined in terms of the selected source reflection coefficient, Γ_S, as given by the equation:

$$\Gamma_L = \left(S_{22} + \frac{S_{12}S_{21}\Gamma_S}{1 - S_{11}\Gamma_S} \right)^*$$

This equation can be solved by letting $\Gamma_S = \Gamma_{opt}$ and calculating the load reflection coefficient, Γ_L.

To use the impedance matching utility we need the conjugate of Z_L or Z_{OUT} of the device or in this case: $Z_{OUT} = 71.9 - j54.4\ \Omega$. Check the location of Γ_L with respect to the output stability circle to make sure that the impedance does not lie in the region of instability. Figure 7-35 shows the location of Γ_L with respect to the output stability circle, SB2.

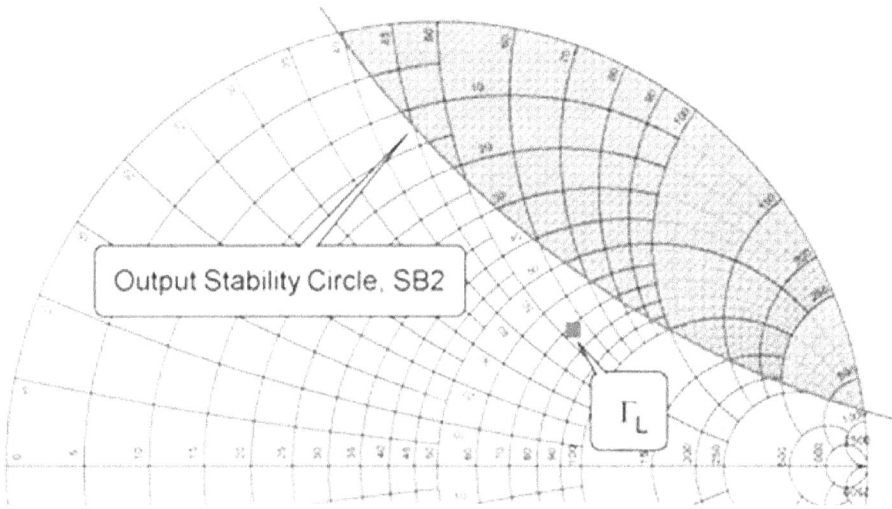

Figure 7-35 Location of Γ_L and the output stability circle

The impedance corresponding to Γ_L can be entered into an impedance element and plotted on the same graph as the output stability circle. The inside of the output stability circle is shaded which means that this area represents unstable output load reflection coefficients. Because Γ_L is safely outside of the SB2 circle we can use Γ_L as an acceptable output reflection coefficient.

Example 7-12: Design a single-stub output matching network for the low noise amplifier using AT30511 device at 432 MHz.

Solution: Now we can design a single-stub output impedance matching network. The procedure follows.

The output load reflection coefficient, Γ_L, is defined in terms of the selected source reflection coefficient, Γ_S. The Genesys Equation Editor is used to calculate the load reflection coefficient, Γ_L.

$$\Gamma_L = \left(S_{22} + \frac{S_{12}S_{21}\Gamma_S}{1 - S_{11}\Gamma_S} \right)^*$$

Where, $\Gamma_S = \Gamma_{opt}$.

```
1.    'Obtain Device S-Parameters from Linear1 Dataset
2     Num=Linear1_Data.S(1,2).*Linear1_Data.S(2,1).*Linear1_Data.GOPT
3     Denom=1-(Linear1_Data).S(1.1).*Linear1_Data.GOPT)
4     GL=Linear1_Data.S(2,2).+(Num./Denom)
5     GL=conj(GL)
6     GL_mag=mag(GL(1))
7     GL_ang=ang(GL(1))*(180/PI)
8     'Calculate Load Impedance
9     ZL=gammatoz(50,ZL)
10    Zload_real=real(ZL(1))
11.   Zload_imag=imag(ZL(1))
```

Figure 7-36 Equation Editor solution of Γ_L and Z_L

To use the impedance matching utility we need the conjugate of Z_L or Z_{OUT} of the device or in this case: $Z_{OUT} = 71.9 - j54.4\ \Omega$. Check the location of Γ_L with respect to the output stability circle to make sure that the impedance does not lie in the region of instability. Figure 7-37 shows the location of Γ_L with respect to the output stability circle, SB2. The impedance corresponding to Γ_L can be entered into an impedance element and plotted on the same graph as the output stability circle. The inside of the output stability circle is shaded which means that this area represents unstable output load reflection coefficients. Because Γ_L is safely outside of the SB2 circle we can use Γ_L as an acceptable output reflection coefficient.

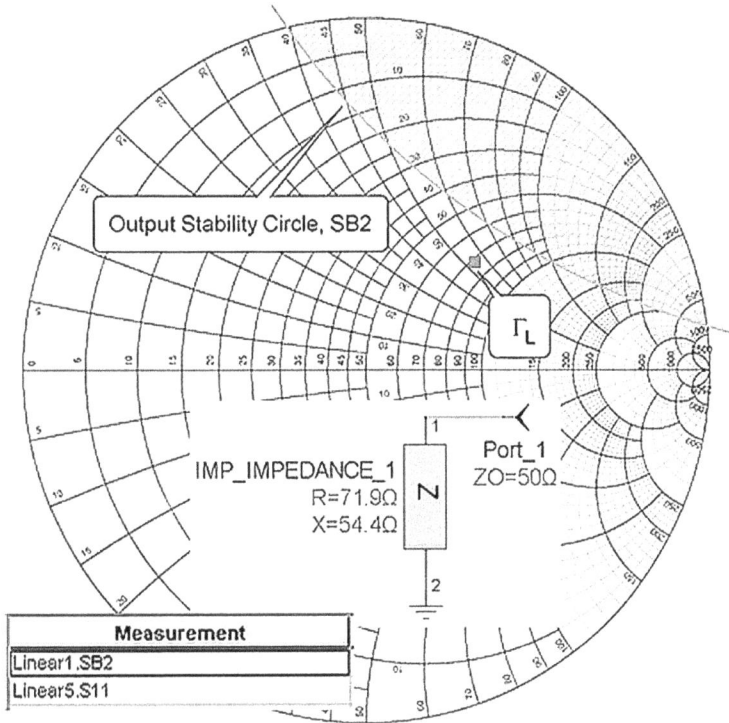

Figure 7-37 Output stability circle and Γ_L

RS=50; RL=71.9; XL=-54.4; f=432e6, r=RL/RS, x=XL/RS

t1 = (x+sqrt(r*(r^2+x^2-2*r+1)))/(r-1)

t2 = (x-sqrt(r*(r^2+x^2-2*r+1)))/(r-1)

d1 = 360*(atan(t1))/(2*pi)

d2 = 360*(pi+atan(t2))/(2*pi)

B1 = (x*t1^2+(r^2+x^2-1)*t1-x)/(RS*(r^2+x^2+t1^2+2*x*t1))

B2 = (x*t2^2+(r^2+x^2-1)*t2-x)/(RS*(r^2+x^2+t2^2+2*x*t2))

so1 = 360*(pi-atan(RS*B1))/(2*pi)

so2 = -360*atan(RS*B2)/(2*pi)

Figure 7-38 Calculation of the single-stub output matching network

The calculations show two single-stub matching network that matches the 50 Ω load impedance to the LNA output impedance. For the output matching network, we select the shorter open-stub length so2 = 44.364 degrees and the corresponding line d2 = 99.959 degrees.

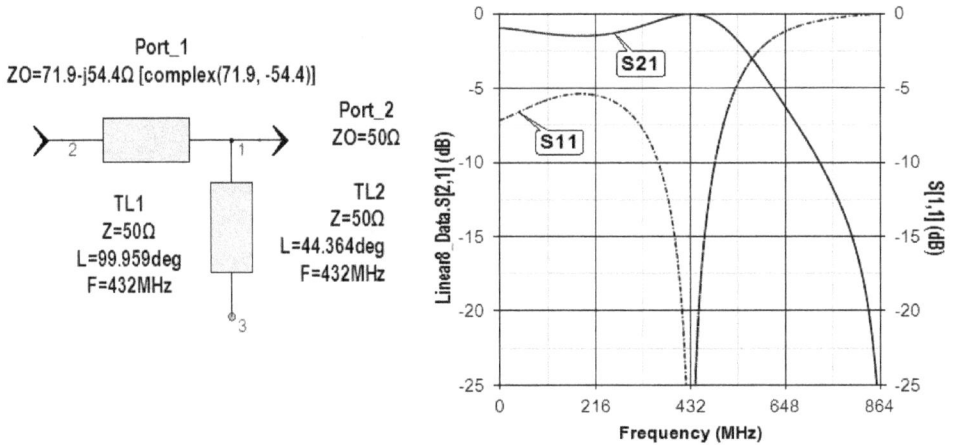

Figure 7-39 Output matching network and simulated response

7.16 Simulation of the Low Noise Amplifier

Example 7-13: Assemble and simulate the distributed element low noise Amplifier.

Solution: Create a new schematic in Genesys with the input and output matching networks attached to the device. Add a new Linear Analysis that sweeps the amplifier from 400 MHz to 500 MHz.

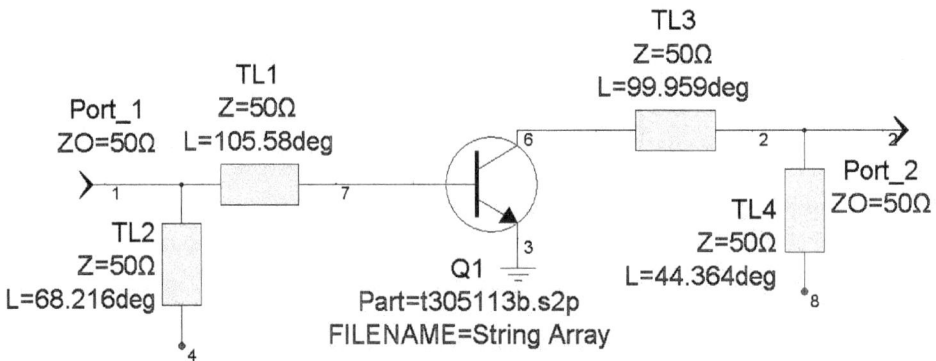

Figure 7-40 LNA schematic using single-stub matching networks

Example 7-14: Assemble the lumped element LNA and measure the amplifier noise figure NF and gain at 432 MHz..

As shown in Figure 7-41, the amplifier gain at 432 MHz is 21.823 dB which is well within the acceptable range of 21.9 dB. The simulated noise figure is 1.083 dB which is exactly the NFmin that was predicted. The remaining design sequence should include the microstrip interconnecting lines and bias feeds. The circuit should then be optimized to bring the final response as close as possible to the ideal circuit results.

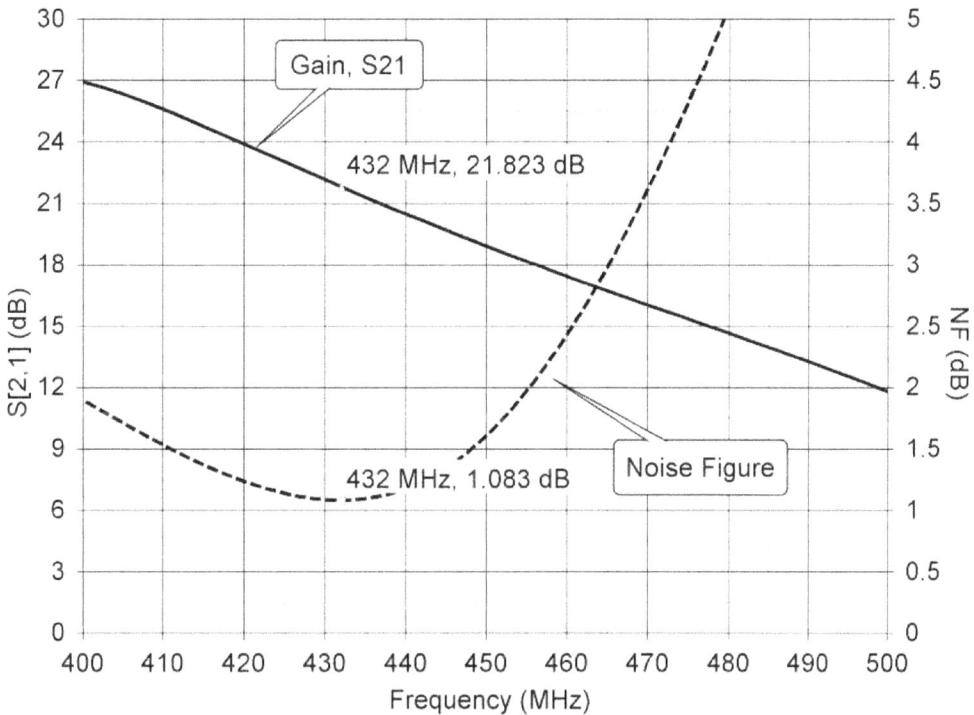

Figure 7-41 LNA noise figure and gain

Figure 7-42 shows that the LNA input is selectively mismatched to achieve the minimum noise figure while the output is perfectively matched to achieve maximum gain. The LNA gain at 432 MHz is 21.823 dB.

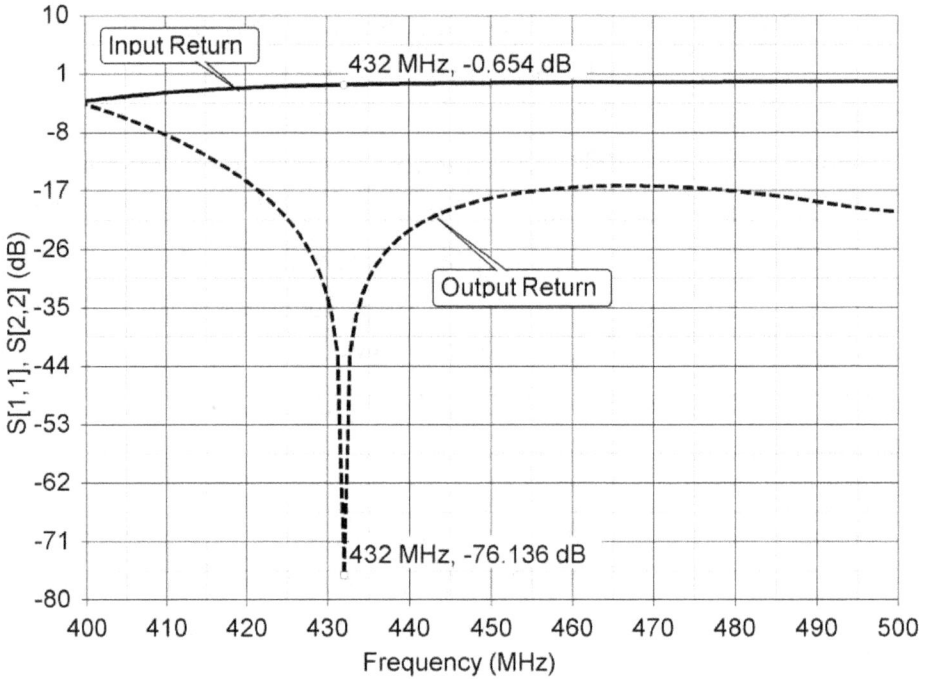

Figure 7-42 LNA input and output return loss

7.17 Power Amplifier Design

The amplifier circuits described thus far have been designed using the measured S parameters that represent the active device. The S parameters are based on the small signal characteristics of the device. That is to say that the device is operating well below its maximum output power capability. This would suggest that the S parameters are defined for class A amplification, operating well within the linear portion of their power transfer characteristic. The S parameters can therefore be used to design an amplifier for specific values of gain but not output power. As we drive the transistor closer to its maximum output power the signal excursion is occurring over a much wider range of the transistor's load line. This is to say that the device is now operating under large signal conditions. There exists a specific source and load impedance into which the transistor can produce its maximum output power. Because the device S parameters are no longer defined under large signal conditions, they cannot be used to determine the optimum load impedance for maximum output power. These impedances are normally determined by performing a load pull measurement on the device.

7.18 Data Sheet Large Signal Impedance

In many cases high power devices are intended for specific applications such as WiMAX, cellular base station, or mobile radio. In these applications the device manufacturer may perform the load pull analysis at specific frequencies and present the source and load impedance as an equivalent series circuit on the data sheet. This allows the designer to treat the power matching process as a simple impedance matching exercise. Figure 7-43 shows an excerpt of a data sheet for the Nitronex NPT25100 GaN HEMT device. As the data sheet shows the optimum source and load impedances are given in tabular and Smith Chart forms.

Frequency (MHz)	Z_S (Ω)	Z_L (Ω)
2140	12.1 - j20.0	2.6 - j2.6
2300	10.0 - j3.0	2.5 - j2.3
2400	9.5 - j3.0	2.5 - j2.5
2500	9.0 - j3.0	2.5 - j2.7
2600	8.5 - j3.0	2.5 - j3.1
2700	8.0 - j3.0	2.5 - j3.3

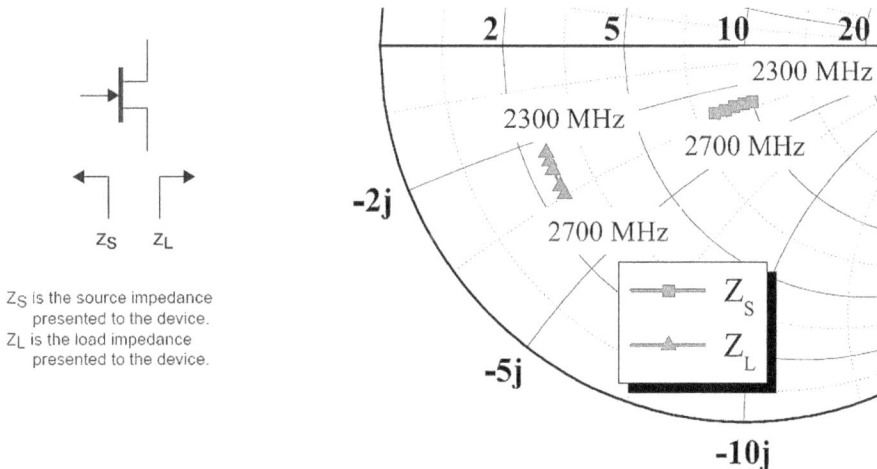

Z_S is the source impedance presented to the device.
Z_L is the load impedance presented to the device.

Figure 7-43 Optimum source and load impedance (*courtesy of Nitronex*)

As Figure 7-43 shows the listed Z_S and Z_L is not the device impedance but rather the impedance that is presented to the device. This is analogous to the Γ_S and Γ_L that was calculated in section 7.2 for the simultaneous conjugate match. The actual device impedance is the conjugate of the given optimum source and load impedance. The designer must be cautious when interpreting the optimum source and load impedances from various vendor data sheets. Some manufacturers may list the actual device impedance as shown on the Freescale Semiconductor data sheet of Figure 7-44. We also need to read the impedance data from the tables rather than directly from the Smith Chart. Because of the very low input and output impedance of power transistors, it is common to normalize the Smith Chart to 10 Ω so that the impedance locus is not compressed on the left hand side of the Smith Chart. The impedances given in the table are the actual impedance rather than the normalized impedance.

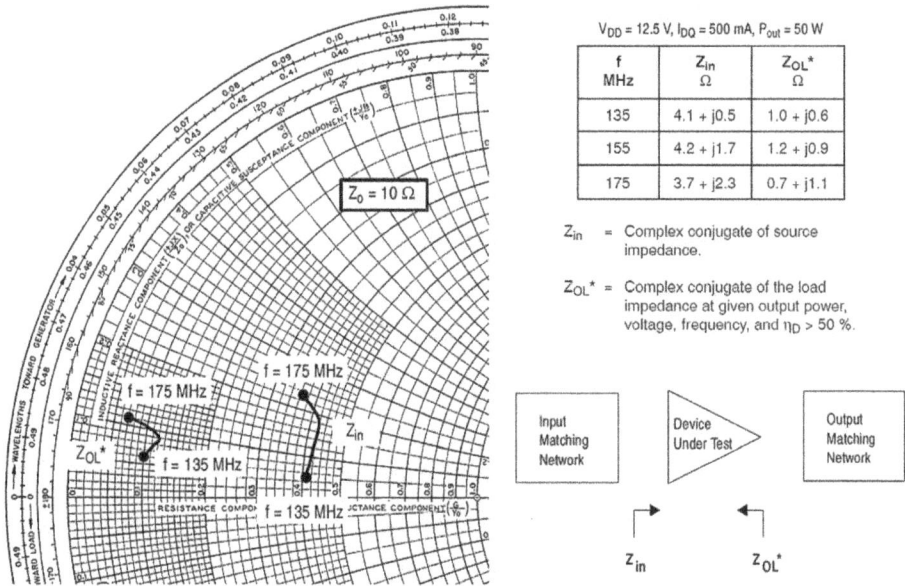

$V_{DD} = 12.5$ V, $I_{DQ} = 500$ mA, $P_{out} = 50$ W

f MHz	Z_{in} Ω	$Z_{OL}*$ Ω
135	4.1 + j0.5	1.0 + j0.6
155	4.2 + j1.7	1.2 + j0.9
175	3.7 + j2.3	0.7 + j1.1

Z_{in} = Complex conjugate of source impedance.

$Z_{OL}*$ = Complex conjugate of the load impedance at given output power, voltage, frequency, and $\eta_D > 50$ %.

Figure 7-44 Large-signal series equivalent impedance (*courtesy of Freescale*)

7.19 Power Amplifier Matching Network Design

In this section we will design the matching networks for the NPT25100 GaN power transistor at a frequency of 2.140 GHz. The matching networks of previous examples have all been based on two-element L-networks. In this example we will design three-element Pi network matching circuits. The Pi network is often preferred over a two element network because the component values are less sensitive when physically realizing the impedance match. It can also help to keep the circuit Q lower for improved bandwidth. We will illustrate both graphical techniques and synthesis techniques. The input matching circuit will be designed using graphical techniques while the output matching network will be designed using network synthesis techniques.

7.20 Power Amplifier Input Matching Network Design

Example 7-15: Design the power amplifier input matching network.

Solution: Graphical matching techniques will be employed for the input matching network using the Smith Chart in Genesys. We will design the matching circuit moving from 50 Ω at the center of the Smith Chart to the Z_S as given on the data sheet. A one port S parameter file can be created with impedance data as shown in Figure 7-45. Note the differences in line number 2 of the file. This line defines the format of the data contained within the file. The (Z) defines the file as containing impedance data. The number (1) means that the data is normalized to one, the actual device impedance.

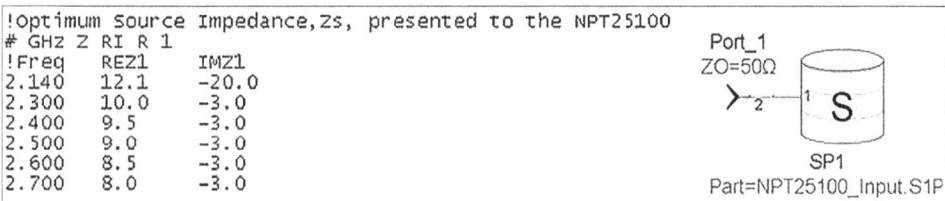

```
!optimum source Impedance,Zs, presented to the NPT25100
# GHz Z RI R 1                                          Port_1
!Freq    REZ1      IMZ1                                  ZO=50Ω
2.140    12.1     -20.0
2.300    10.0      -3.0
2.400     9.5      -3.0
2.500     9.0      -3.0                                      S
2.600     8.5      -3.0
2.700     8.0      -3.0                                  SP1
                                            Part=NPT25100_Input.S1P
```

Figure 7-45 One port S parameter data file containing the Z_S data

Setup a schematic in Genesys with the one-port Z_S impedance data and sweep it with a Linear Analysis at a fixed frequency of 2140 MHz. Then setup a second schematic with a 50 Ω resistor to represent the center of the Smith Chart and begin the Pi network design. Plot the impedance of both networks on the same Smith Chart. Figure 7-46 through 7-48 show the progression of the three-element Pi matching network design. We know that the shunt inductors of a Pi network will travel clockwise on the conductance circles while the series inductor will travel clockwise on a resistance circle. Note the location of the conductance circle in which Z_S is located. The first capacitor and series inductor L section must move to the location of the Z_S conductance circle.

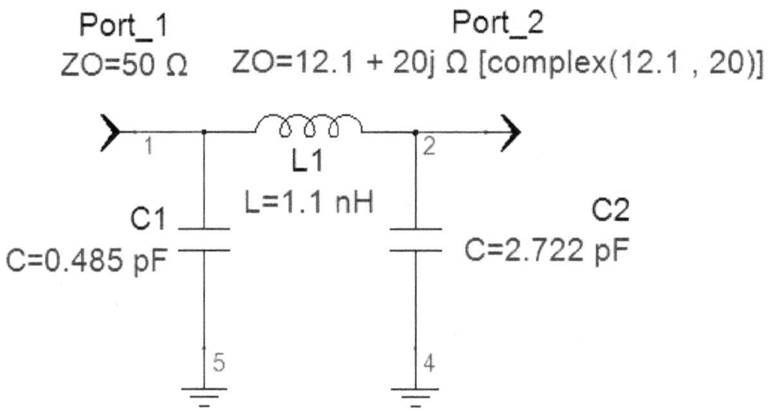

Figure 7-46: Power amplifier input matching circuit

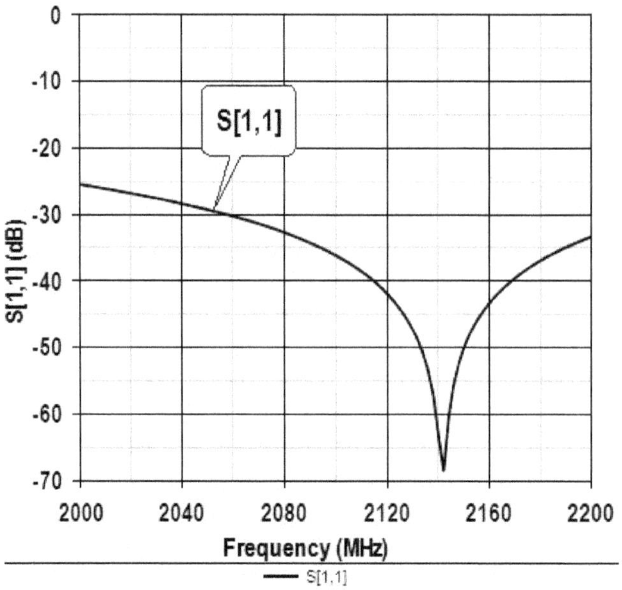

Figure 7-47: Input Matching Response.

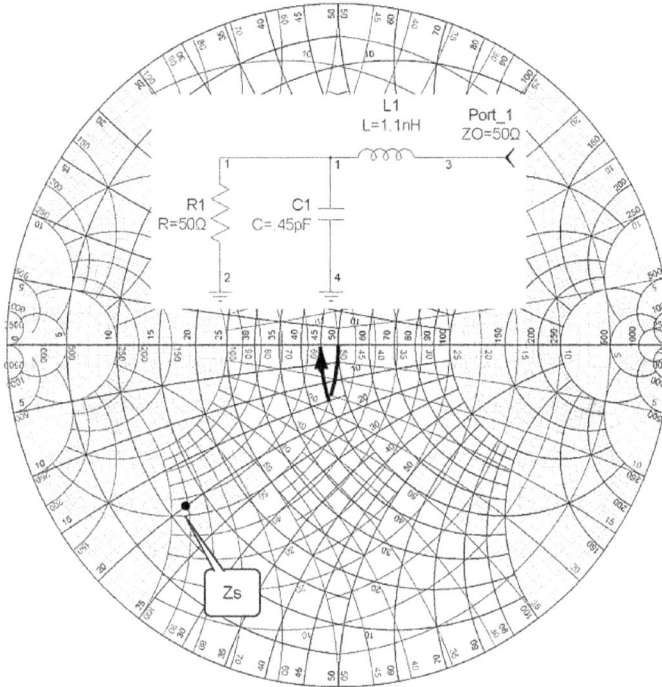

Figure 7-47 Tuning the series inductor to intercept the conductance circle

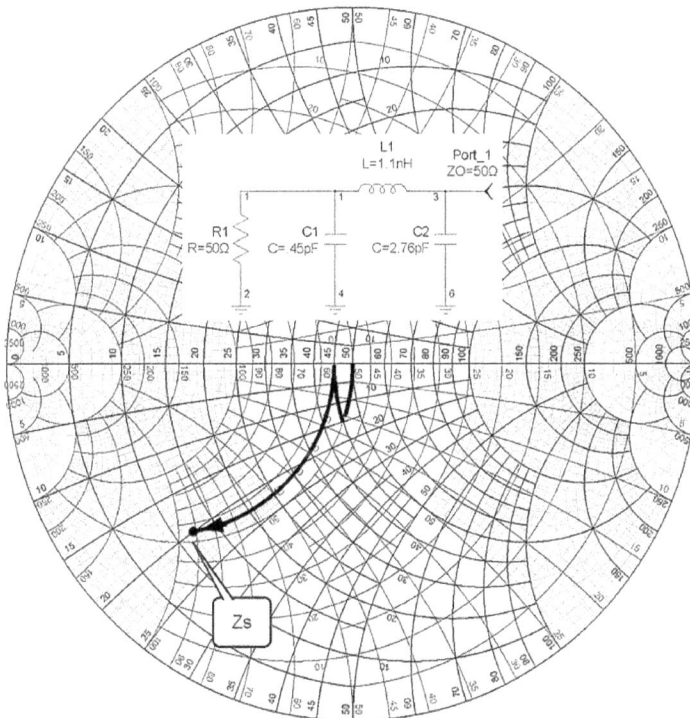

Figure 7-48 Tuning the second shunt capacitor to intersect Z_S

7.21 Power Amplifier Output Matching Network Design

Example 7-16: Design a three elements Pi network to match the 50 Ω load impedance to the amplifier output at 2140 MHz.

Solution: The output matching network will be designed using the Impedance Matching Synthesis utility. When using the synthesis utility we need to have the device impedance rather than the impedance looking into the matching network toward the load. Therefore a one port S parameter file is created with the conjugate of the Z_L given on the data sheet as shown in Figure 7-49. Use caution when editing S parameter files to be read into a Genesys simulation. The first time that an S parameter file is read from disk it is loaded into the Genesys Workspace. Subsequent simulations will not go back to disk to read the file but rather read the file from the Workspace for faster simulation speed. The file will not be read from disk again until the copy loaded into the Workspace has been deleted. Create an Impedance Match Synthesis for the output matching network. Set the properties as shown in Figure 7-50. Enter a fixed frequency of 2140 MHz.

On the Sections tab, read the one-port S parameter file that contains the conjugate of the Z_L from the data sheet. Next select an LC Pi Network for the matching circuit with an inductive tendency. Select Calculate to synthesize the matching network shown in Figure 7-51.

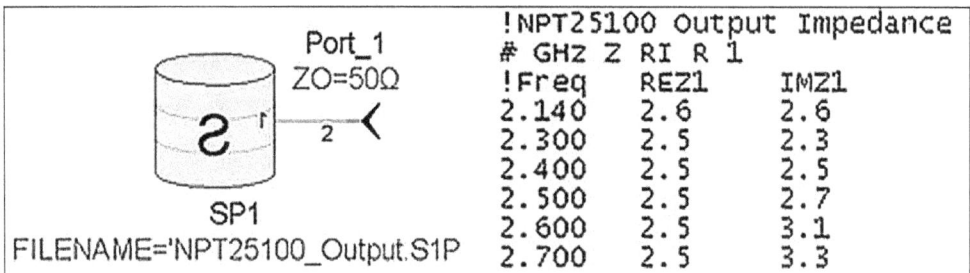

!Freq	REZ1	IMZ1
2.140	2.6	2.6
2.300	2.5	2.3
2.400	2.5	2.5
2.500	2.5	2.7
2.600	2.5	3.1
2.700	2.5	3.3

!NPT25100 Output Impedance
GHZ Z RI R 1

Port_1
ZO=50Ω

SP1
FILENAME='NPT25100_Output.S1P

Figure 7-49 S parameter data file containing the conjugate of Z_L

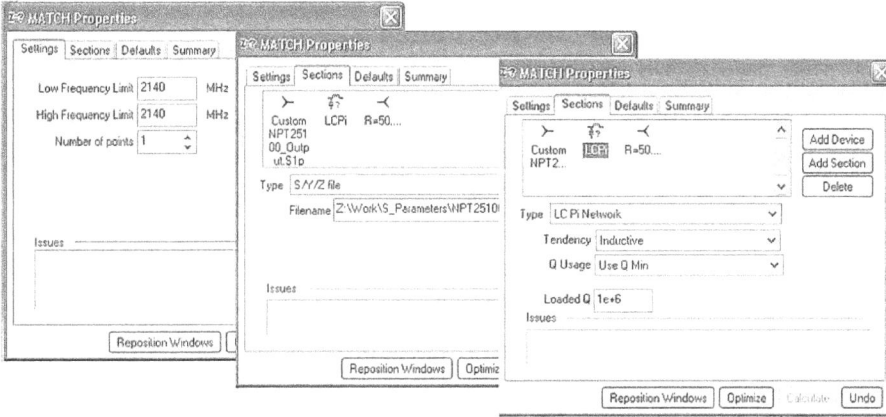

Figure 7-50 Setup of the output matching network synthesis parameters

The impedance matching synthesis program is an excellent tool for quickly examining various matching networks and their response.

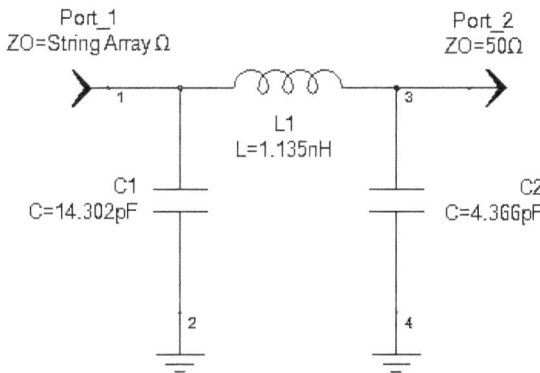

Figure 7-51 Output matching network

Finally a two-port S parameter file can be created that contains the impedance data for the conjugate of Z_S and Z_L. This would be representative of the large signal input and output impedance of the device.

```
'Large Signal Impedances of the NPT25100
# GHz Z RI R 1
'Freq      REZ1   IMZ1      M(S21)    A(S21)     M(S12)    A(S12)     REZ2   IMZ2
2.140      12.1   20.0      0         0          0         0          2.6    2.5
2.300      10.0   3.0       0         0          0         0          2.5    2.3
3.400      9.5    3.0       0         0          0         0          2.5    2.5
2.500      9.0    3.0       0         0          0         0          2.5    2.7
2.600      8.5    3.0       0         0          0         0          2.5    3.1
2.700      8.0    3.0       0         0          0         0          2.5    3.3
```

Figure 7-52 S parameter data file containing the conjugate of Z_S and Z_L

Figure 7-53 shows the resulting matching networks attached to the large signal impedance data for the device. This allows examination of the return loss of the amplifier at the design frequency of 2140 MHz and the usable bandwidth. From Figure 7-53 note that the forward (S21) and reverse (S12) transmission parameters have been set to zero. Because we have no definition of the transmission parameters we cannot evaluate the gain of the circuit. If the manufacturer's data sheet includes an S parameter file along with the large signal impedance data, we could also examine the stability parameters to determine whether the device may require a stabilization network. The Linear Design techniques do provide a means of performing the matching network design from which the amplifier physical design can be realized. The circuit can then be built and empirically tuned and optimized on the bench for the desired performance. A thorough CAD design of the power amplifier requires a nonlinear physical model for the transistor. Under large signal conditions the strong nonlinearities will cause the gain to change as the drive (input) power changes. The gain of the device will go into compression and decrease as the input power is further increased. Harmonic energy is created by these nonlinearities that will further influence the behavior of the amplifier.

Example 7-17: Assemble the power amplifier, simulate, and display the response from 200 to 2200 MHz.

Solution: Create a new schematic in Genesys with the input and output matching networks attached to the device. Add a new Linear Analysis that sweeps the amplifier from 2 to 2.2 GHz.

SP1
FILENAME='NPT25100.S2P'

Port_1
ZO=50Ω

L2
L=1.1nH

L1
L=1.135nH

Port_2
ZO=50Ω

C4
C=.45pF

C3
C=2.76pF

C1
C=14.302pF

C2
C=4.366pF

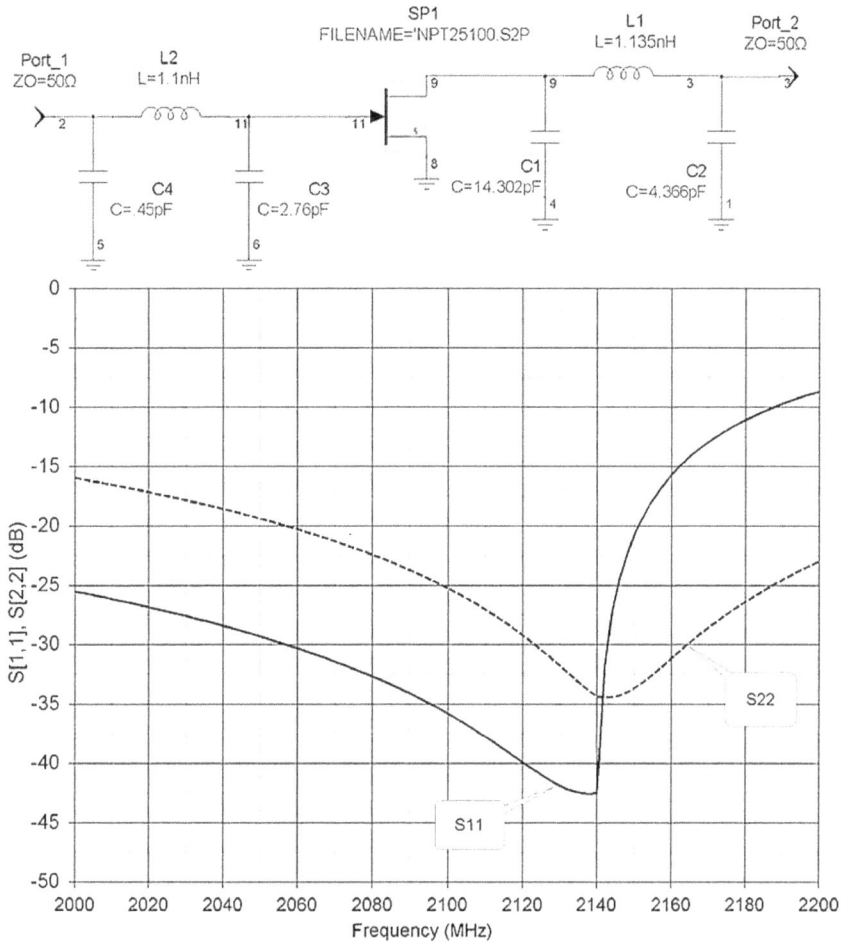

Figure 7-53 Resulting matching networks and return loss

The impedance matching techniques learned in Chapters five and six have been utilized to perform basic linear amplifier matching. These matching techniques have been solved analytically using the equations written in the Genesys Equation Editor as well as synthesis routines created in VBScript. Additionally the powerful matching synthesis tools that exist within Genesys have also been applied. Graphical techniques have also been applied to the design of multi-element impedance matching using the Smith Chart. These techniques provide the engineer with a comprehensive set of tools to apply to amplifier impedance matching. These impedance matching techniques have been used to introduce four of the primary RF and microwave amplifier circuits including: maximum gain, specific gain, low noise, and power amplifier circuits.

References and Further Readings

[1]   Dale D. Henkes, FAST: *Fast Amplifier Synthesis Tool*, Artech House Publishers, Norwood, MA. 2004

[2]   Chris Bowick, *RF Circuit Design*, Butterworth-Heinemann, Newton, MA, 1982

[3]   Keysight Technologies, *Genesys 2015.08, Users Guide*, 2015 www.keysight.com

[4]   Guillermo Gonzales, *Microwave Transistor Amplifiers – Analysis and Design*, Second Edition, Prentice Hall Inc., New Jersey

[5]   Randy Rhea, *The Yin-Yang of Matching: Part 1 – Basic Matching Concepts*, High Frequency Electronics, March 2006

[6]   Steve C. Cripps, *RF Power Amplifiers for Wireless Communications*, Artech House Publishers, Norwood, MA. 1999

[7]   David M. Pozar, *Microwave Engineering*, Fourth Edition, John Wiley & Sons, New York, 2012

[8]   R. Ludwig, P. Bretchko, *RF Circuit Design*, Theory and Applications, Prentice Hall, Upper Saddle River, NJ, 2000

[9]   Ali A. Behagi and Stephen D. Turner, *Microwave and RF Engineering,* A Simulation Approach with Keysight Genesys Software, March, 2015

Problems

7-1. Design a Maximum Gain Amplifier

(a) For the Agilent HBFP0405 Transistor, calculate the stability factor, K, and the stability measure, B1 as well as Γ_{MS} and Γ_{ML} and conjugate match impedances at 2 GHz.

(b) Use Genesys built in functions to determine the same parameters at 2 GHz.

(c) Sweep the frequency range of the device over the entire range of frequencies contained in the S parameter file.

(d) Plot of the stability parameters and GMAX for the Agilent HBFP0405 device from 100 to 4000 MHz.

(e) Employ a parallel RC circuit on the input of the device to stabilize the transistor. Make both the R and C values tunable. Tune the resistance and capacitor values until K > 1 for frequencies above 500 MHz.

(f) Add an RL network in shunt with the input of the transistor for additional low frequency stability. Tune the resistance and inductor values until K > 1 for frequencies above 500 MHz.

(g) Use the analytical impedance matching techniques developed in Chapter 5 to design the input and output matching L-networks. When performing the analytical match we are; matching 50 Ω source impedance to the impedance looking into the device; and 50 Ω load impedance to the impedance looking into output of the device.

(h) Complete the amplifier design in Genesys by attaching the input and output matching circuits to the stabilized transistor. Use a Linear Sweep from 100 MHz to 4000 MHz to analyze the response of the amplifier, S11, S22, and S21 all in dB.

(i) Place a marker on S12 to measure the maximum gain of the amplifier at 2 GHz. Compare the maximum gain with the value originally calculated as G_{MAX}.

7-2. Design a Constant Gain Amplifier

(a) Use the Mitsubishi MGF0911a device to design a specific gain amplifier that achieves 10 dB gain at the frequency of 2 GHz. Examine the stability circles for the device in Genesys by creating a schematic with the small signal S parameter file.

(b) Setup a Linear Analysis with both the start and stop frequencies set at 2 GHz. Add a Linear Analysis from 100 MHz to 3000 MHz with 100 MHz steps. Then add a Table to display K, B1, and GMAX.

(c) Plot a series of constant power gain circles on the Smith Chart and choose the circle which is the closest to the design goal of 10 dB. The load reflection coefficient, defined as a reflection coefficient, Γ_L, is any point on the circumference of the circle that does not enter the unstable region. Choose a point that is not close to the unstable regions of the stability circles. Read the load impedance visually from the Smith Chart at this point.

(d) Calculate the source impedance from Equation 7-18. The source impedance is therefore defined as a function of the chosen load reflection coefficient.

(e) Use the Smith Chart to graphically design the simple two-element LC input and output matching networks.

(f) Create a new schematic and attach the input and output matching circuits to the device. Create a Linear Analysis that sweeps the amplifier from 1 GHz to 3 GHz. Display S11 and S21 in a rectangular plot.

(g) Place a marker on S12 to measure the gain of the amplifier at 2 GHz. Compare the gain with the value originally selected as specific gain.

7-3. Design a Low Noise Amplifier

Design a single stage Low Noise Amplifier using Agilent AT-32011 device. The amplifier is intended to operate with a source and load impedance of 50 Ω. The design specifications are given as:

Center Frequency:	500 MHz
Gain:	18 dB minimum
Noise Figure:	1.2 dB maximum
Output Return Loss:	Less than -10 dB

(a) Create a Genesys schematic with the device S parameter file. Setup a Linear Analysis and add a Table to display the stability parameters K and B1 and the noise parameters Γ_{opt} and Z_{OPT}. Also add a Smith Chart with Γ_{opt} and the stability circles. Then create a Smith Chart and display the stability circles along with the noise circles.

(b) Drawn the available gain circles are for gains of 0, 1, 2, 3, 4, 5 and 6 dB less than maximum gain.

(c) Utilize the "L" Network Impedance Matching Synthesis Tool to match the 50 Ω source impedance to Z_{opt} of the device at 500 MHz. In the impedance matching utility you must enter the conjugate of Z_{opt} for the complex load impedance.

(d) Use Genesys Equation Editor to calculate the load reflection coefficient, Γ_L, from Equation (7-24). Convert Γ_L to output impedance Z_L. Z_{OUT} is the conjugate of Z_L. Check the location of Γ_L with respect to the output stability circle to make sure that the impedance does not lie in the region of instability. If Γ_L is safely outside of the SB2 circle, use Γ_L as an acceptable output reflection coefficient.

(This page is intentionally left blank)

Multi-Stage Amplifier Design

8.1 Introduction

Practical amplifiers often involve a cascade of amplifier stages to provide the required gain for a specific application. In an amplifier cascade, the output impedance of one device is typically matched to the input impedance of the succeeding stage rather than a 50 Ω resistance. This chapter provides an overview of amplifier inter-stage matching. For simplicity much of this chapter will focus on ideal schematic elements for both inter-stage matching and the Yield Analysis. Finally a brief introduction to cascade analysis is presented as it can be applied to linear simulation in Genesys.

8.2 Two-Stage Amplifier Design

Figure 8-1 shows a block diagram of a two stage amplifier with the matching networks represented by a box. The input impedance of each device is designated as ZM1 while the output is designated as ZM2. We will revisit two of the transistors used in Chapter 7 to design a two stage amplifier cascade. The first stage is the SHF0189 and the second stage is the RT243. The design specifications are given as:

Center Frequency = 2350 MHz
Bandwidth: 2260 MHz to 2380 MHz
Gain: ≥ 34 dB   −
Input & Output Return Loss < -10 dB

The cascade is designed as a maximum gain-conjugate matched amplifier. Therefore three conjugate matching networks must be designed. The first conjugate matching network is designed between the 50 Ω source and the first stage input impedance, ZM1. The second matching network is the inter-stage matching network. This network is designed to match the first stage output impedance, ZM2, to the second stage input impedance, ZM1.

The third conjugate matching network is designed between the second stage output impedance, ZM2, and the 50 Ω load impedance.

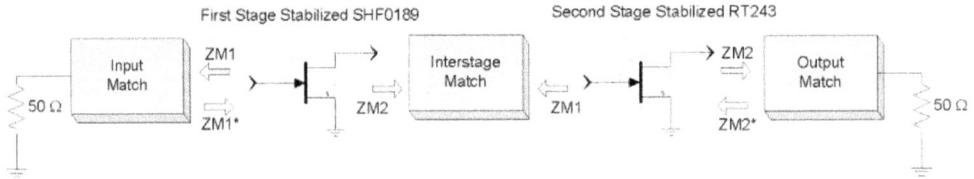

Figure 8-1: Two-stage amplifier with impedance matching networks

8.3 First Stage Stability Considerations

The first stage of the amplifier utilizes the SHF-0189 device. We know that the device must be unconditionally stable for the maximum gain conjugate matched amplifier. Using the techniques of Chapter 7 a stabilizing network is designed around the SHF-0189.

Example 8-1 Measure the input-output impedance and the maximum stable gain of the first stage of the cascaded amplifier by utilizing the SHF-0189 device at 2350 MHz.

Solution: Once the device is stabilized the maximum stable gain and the required input and output simultaneous match impedance is determined by using the built-in functions in Genesys. Create a new workspace and a new schematic in Genesys. Use the stabilizing network developed in Chapter 7 for the SHF-0189 device as shown in Figure 8-2.

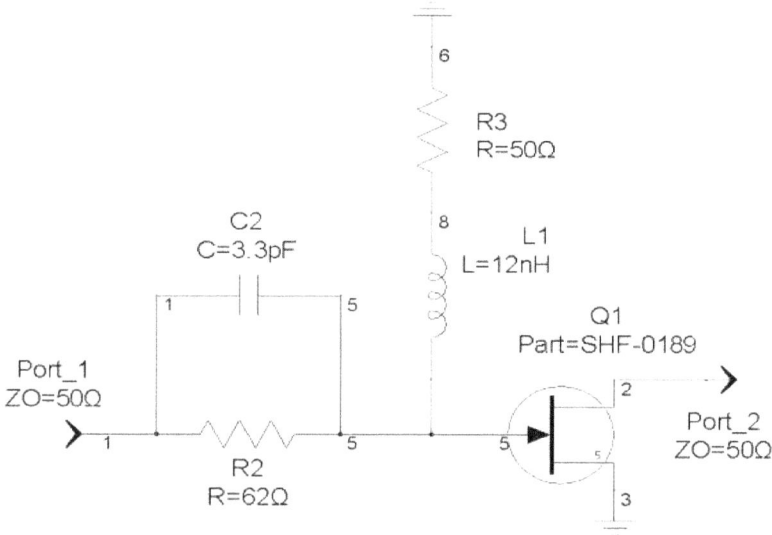

F (MHz)	GMAX (dB10)	re(ZM1)	im(ZM1)	re(ZM2)	im(ZM2)
2350	15.779	6.947	22.058	13.773	16.471

Figure 8-2 Stabilized first stage with conjugate match impedance

The Genesys input-output impedances are shown in the table inset of Figure 8-2. The equivalence between the Genesys function notation and the respective impedances is given below.

$$ZM1 = Z_{MS} = 6.947 + j\,22.058 \ \Omega$$

$$ZM\,2 \ = \ Z_{ML} \ = \ 13.773 + \ j\,16.471 \ \Omega$$

$$Gmax = 15.779 \ \ dB$$

8.4 Amplifier Input Matching Network Design

Example 8-2: Design the Amplifier Input Matching Network

Solution: When designing the amplifier input matching network we are matching the 50 Ω source impedance to the conjugate impedance looking into the SHF-0189 device, ZM1, as given in Figure 8-2. Therefore the matching of the complex impedance is the conjugate of ZM1 as written in the following Equation editor. Normalize the load impedance and use Equations (5-32) and (5-33) to calculate the matching element values.

```
 2    Z0=50
 3    RL=6.947
 4    XL=-22.058
 5    f=2.35e9
 6    r=RL/Z0
 7    x=XL/Z0
 8
 9    B1=sqrt((1-r)/r)/Z0
10    X1=Z0*(sqrt(r*(1-r))-x)
11    L1=X1/(2*pi*f)
12    C2=B1/(2*pi*f)
```

Figure 8-3 Calculation of the element values of the input matching network

The calculations in the Equation Editor show that the series element is an inductor L1= 2.665 nH and the shunt element is a capacitor, C2 = 3.372 pF. To assemble the schematic, display the response and measure the bandwidth, set up a new Schematic and select the element values given above. Simulate the schematic from 1350 MHz to 3350 MHz and display the insertion loss, S21, and input return loss, S11, in dB. The schematic of the matching network and the simulated response are shown in Figure 8-4. Put two markers at 20 dB points to measure the bandwidth as shown in Figure 8-4. The SHF-0189 output matching network is not designed here because we want to cascade the SHF-0189 with the RT243 device and design an inter-stage matching network between the two stages.

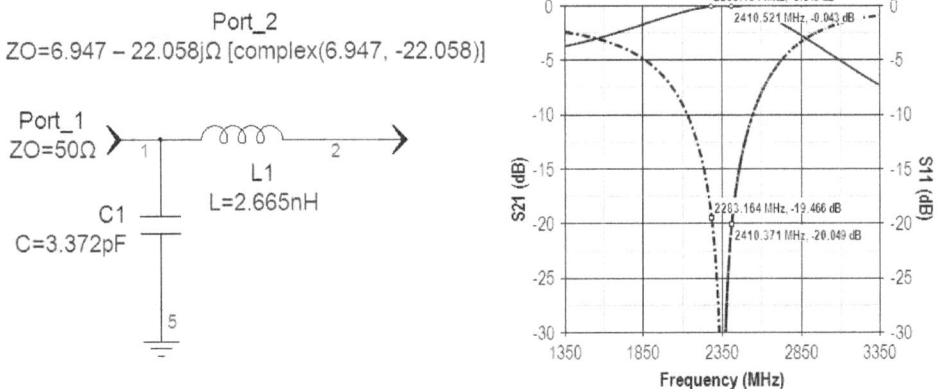

Figure 8-4 Schematic of the input matching network and response

8.5 Second Stage Matching Network Design

Example 8-3: Measure the simultaneous match input-output impedance and the maximum stable gain of the second stage of the cascaded amplifier by utilizing the RT243 device at 2350 MHz.

Solution: The stabilized second stage device is shown in Figure 8-5. Once the device is stabilized the maximum stable gain and the required input and output impedance are given by the built-in functions in Genesys. Figure 8-5 shows that RT243 has a GMAX = 18.973 dB at 2.35 GHz.

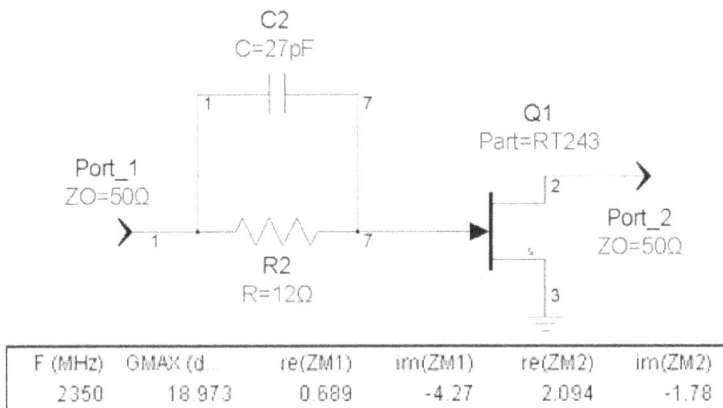

F (MHz)	GMAX (d...	re(ZM1)	im(ZM1)	re(ZM2)	im(ZM2)
2350	18.973	0.689	-4.27	2.094	-1.78

Figure 8-5 Stabilized second stage with conjugate match impedance

The equivalence between the Genesys function notation are:

$$ZM1 = Z_{MS} = 0.68 - j\,4.27\,\Omega$$

$$ZM2 = Z_{ML} = 2.09 - j1.78\,\Omega$$

$$G_{max} = 18.97\ dB$$

8.6 Inter-Stage Matching Network Design

One common method of inter-stage matching is to design each stage separately into a 50 Ω system and then cascade them directly with no additional matching. Many RF transistors have very low input and output impedance making the interstage network desirable. In cascaded narrowband amplifier design a more efficient method of interstage matching is to conjugately match the two complex impedances directly together with a single L-network and thus reduce the number of matching elements.

Therefore, for the interstage matching network we can use the equations developed in Chapter 5 to directly match together the two complex impedances.

Example 8-4: Design the interstage matching network between the SHF-0189 and the RT243 devices.

Solution: The element values of the inter stage matching networks are calculated as follows.

```
RS=13.773
XS=-16.471
RL=0.688
XL=4.271
f=2.35e9

B3=((RS*XL)+sqrt(RS*RL*(RL^2+XL^2-RS*RL)))/(RS*(RL^2+XL^2))
X3=(RS*XL-RL*XS)/RL+(RL-RS)/(B3*RL)
L2=X3/(2*pi*f)
C1=B3/(2*pi*f)
```

Figure 8-6 Calculation of interstage matching network elements

The inter-stage matching network between SHF-0189 and RT243 is shown in Figure 8-7. To display the response of the interstage matching network, set up a new Design with Schematic in Genesys and place the matching elements with the source and load impedances on the schematic, as shown. Simulate the schematic from 1350 MHz to 3350 MHz and display both the return loss, S11, and insertion loss, S21, in dB, as shown in Figure 8-7.

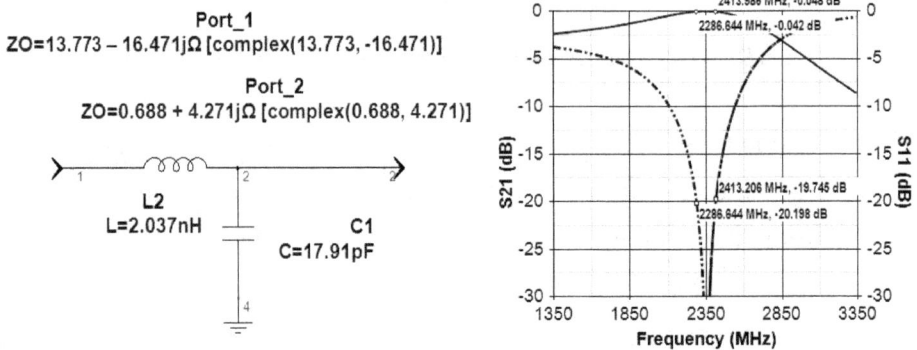

Port_1
ZO=13.773 – 16.471jΩ [complex(13.773, -16.471)]

Port_2
ZO=0.688 + 4.271jΩ [complex(0.688, 4.271)]

L2
L=2.037nH

C1
C=17.91pF

Figure 8-7 Schematic and response of the interstage matching network

8.7 Second Stage Output Matching Network

Example 8-5: Design the second stage output matching network.

Solution: Use Equations (5-24) and (5-25) to calculate the element values of the output matching network. The schematic in Figure 8-8 shows that the shunt capacitor C1=6.48 pF and the series inductor L1 = 0.577 nH. Then set up new schematic in Genesys and select the element values from the Equation Editor for the output matching network. Simulate the schematic from 1350 MHz to 3350 MHz and display the insertion loss, S21, in dB.

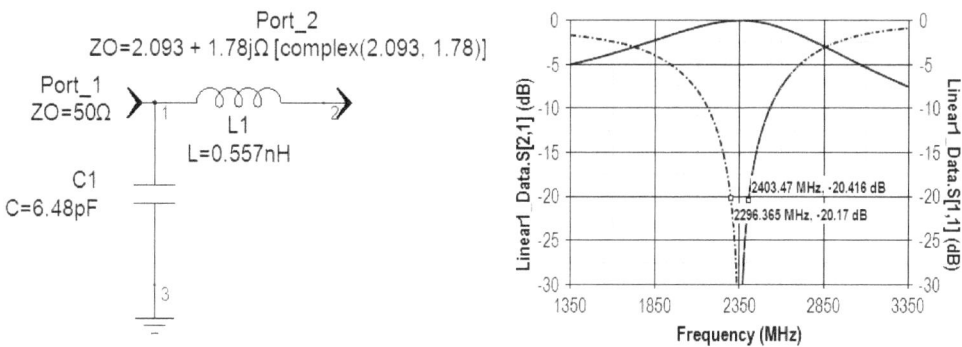

Figure 8-8 Schematic and response of the output matching network

8.8 Two-Stage Amplifier Simulation

Example 8-6: Assemble, simulate, and display the response of the two stage amplifier.

Solution: The matching networks are attached to the stabilized devices to form a two stage amplifier. The input, inter-stage, and output matching networks are cascaded along with the devices as shown in Figure 8-9. We will only deal with the ideal schematic at this point. The final design, of course, would include the physical models of the lumped element components and the interconnecting microstrip circuit elements.

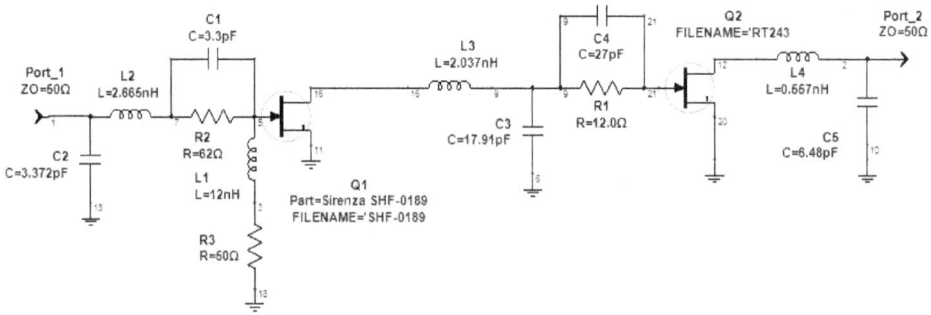

Figure 8-9 Two-stage amplifier schematic with ideal matching networks

The simulated gain and return loss are shown in Figure 8-10. The simulated gain of the two-stage cascade at 2.35 GHz is 34.564 dB which is very close to the direct summation of the GMAX for each stage. The gain of the amplifier is greater than 34 dB from 2260 MHz to 2380 MHz, satisfying one of the design specifications of the amplifier. The input and output return loss is well below -10 dB, thereby satisfying all of the specifications of the amplifier.

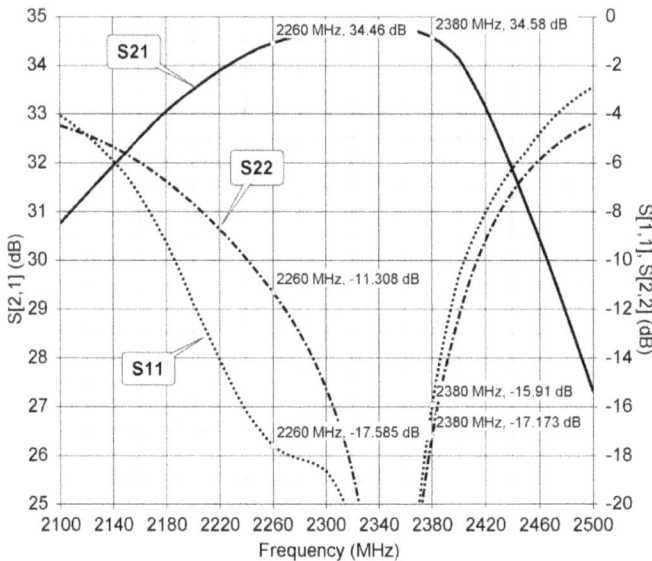

Figure 8-10 Ideal amplifier cascade response

8.9 Multi-Stage Low Noise Amplifier Cascade

In practical amplifier applications, a single stage amplifier is rarely adequate to meet the overall gain and output power requirements of an amplifier design. Therefore it is necessary to cascade multiple gain stages to achieve a particular gain specification. In this section the Gain and Noise Figure of the multi-stage cascaded LNA is discussed.

8.10 Cascaded Gain and Noise Figure

As we have seen in the example of Section 8.3 the overall gain of a cascaded amplifier is simply the algebraic sum of the gains (or losses) in dB. The gain of an amplifier cascade is then given by Equation (8-7).

$$Cascade\ Gain, G_{dB} = G1_{dB} + G2_{dB} + G3_{dB} + ... \qquad (8\text{-}7)$$

The overall noise figure is defined by the Friis formula [6]. As the Friis formula of Equation (8-8) shows, the overall noise figure of a cascade is influenced by all stages in the cascade. The noise figure contribution of the second stage is reduced by the gain of the first stage. Therefore to maintain a low noise figure it is important that the first stage have high gain. The noise factor of a given stage is reduced by the gain of the preceding stages throughout the cascade.

$$Noise\ Factor,\ F\ =\ F_1 + \frac{F_2 - 1}{G_1} + \frac{F_3 - 1}{G_1 G_2} + \frac{F_4 - 1}{G_1 G_2 G_3} + ... \qquad (8\text{-}8)$$

where,

$$Noise\ Factor,\ F = 10^{\frac{F_{dB}}{10}} \qquad\qquad Gain,\ G = 10^{\frac{G_{dB}}{10}}$$

In the solution of the Friis formula it is important to realize that the noise figure and gain must not be in dB format. This is referred to as the noise factor and related to the noise figure by the 10[log(F)] function. For the three-stage LNA of Figure 8-11, the cascaded gain is easily calculated as 41 dB.

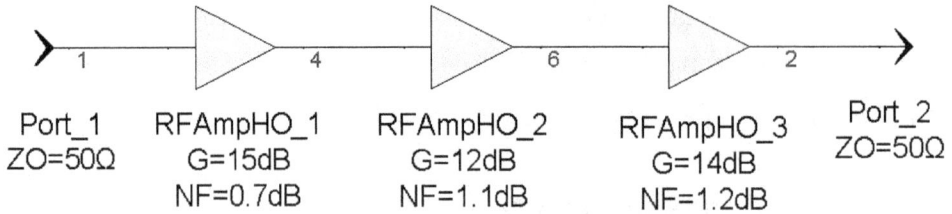

Port_1   RFAmpHO_1   RFAmpHO_2   RFAmpHO_3   Port_2
ZO=50Ω   G=15dB   G=12dB   G=14dB   ZO=50Ω
                 NF=0.7dB   NF=1.1dB   NF=1.2dB

Figure 8-11 Three-stage cascaded low noise amplifier

Using the Friis formula to solve for the resulting noise factor leads to the following result.

$$F = 10^{\frac{0.7}{10}} + \frac{10^{\frac{1.1}{10}} - 1}{10^{\frac{15}{10}}} + \frac{10^{\frac{1.2}{10}} - 1}{10^{\frac{15}{10}} \cdot 10^{\frac{12}{10}}} = 1.1844$$

Converting the noise factor back to noise figure gives a total cascaded noise figure of 0.736 dB.

$$F_{dB} = 10 \cdot \log(1.1844) = 0.736 \ dB$$

The cascaded gain and noise figure can be calculated in Genesys using the simple block diagram of Figure 8-22. This three stage cascade uses the RFAmpHO system level model but can be used in Linear simulation for gain and noise figure.

Setup a Linear simulation for any frequency range as the frequency is independent for this use of the RFAmpHO model. The simulated cascade gain, S21, and noise figure, NF, are shown in Figure 8-12. We can see that the overall noise figure of the cascade is 0.736 dB which correlates with the solution of the Friis formula. The noise figure of the first stage dominates the overall noise figure and is slightly degraded by the 2nd and 3rd stages.

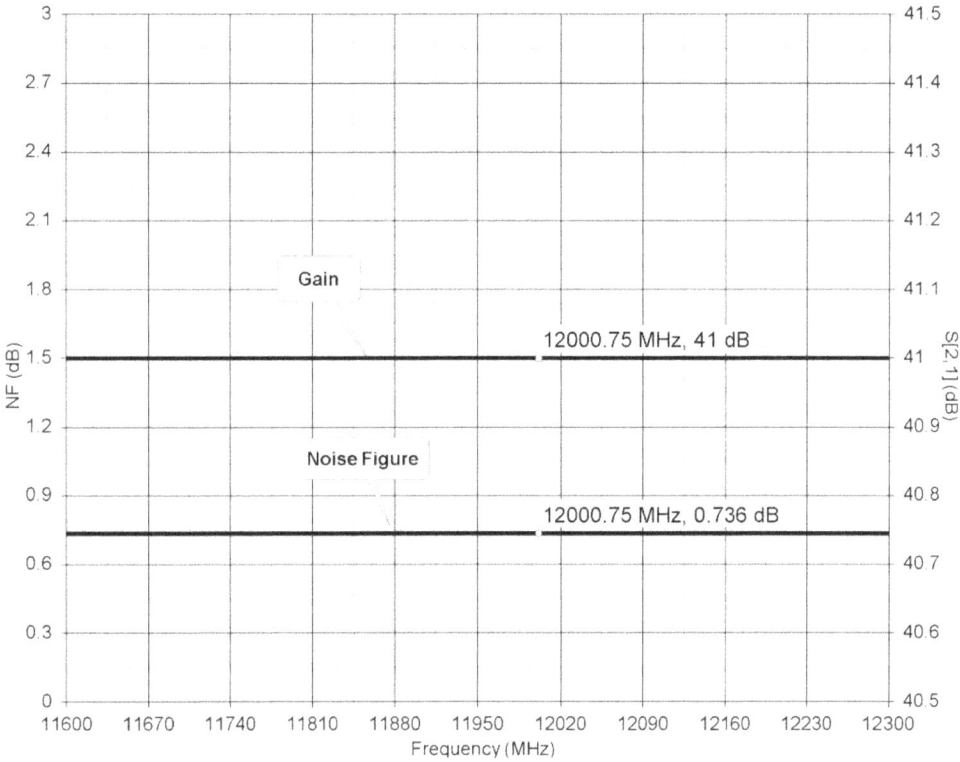

Figure 8-12 Simulated LNA cascade noise figure and gain

8.11 Impedance Match and the Friis Formula

A Low Noise Amplifier is typically used as the front end of a radio receiver. Therefore it is often attached to an antenna and filter combination. From Chapter 4 we have seen that the impedance of a filter can vary significantly across its passband as determined by the ripple and return loss. The same can be true of an antenna or the antenna feed network. Because the Noise Figure of an LNA is dependent on its source impedance, large errors can be obtained in the application of the Friis formula when calculating the overall noise figure of such a cascaded network. The Friis formula assumes that there is a perfect impedance match between each stage in the cascade and there are no variations in impedance across the frequency band.

Example 8-7: Analyze a typical cascade of Low Noise Amplifier used in the range of 11.7 to 12.2 GHz. To eliminate any interference from the uplink signal or terrestrial sources a low loss waveguide filter is placed at the input of the LNA. This system level block diagram is shown in Figure 8-13.

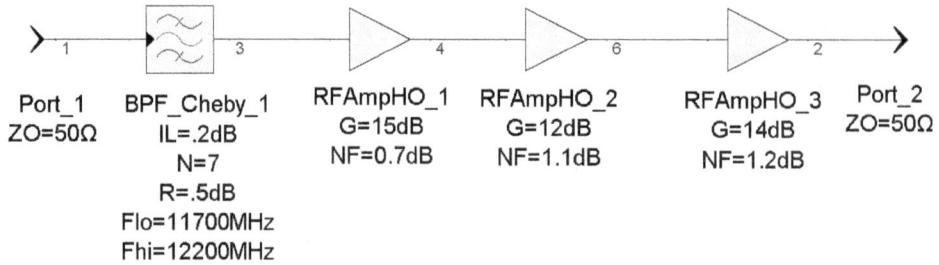

| Port_1 | BPF_Cheby_1 | RFAmpHO_1 | RFAmpHO_2 | | RFAmpHO_3 | Port_2 |

Port_1   BPF_Cheby_1   RFAmpHO_1   RFAmpHO_2   RFAmpHO_3   Port_2
ZO=50Ω   IL=.2dB   G=15dB   G=12dB   G=14dB   ZO=50Ω
                N=7   NF=0.7dB   NF=1.1dB   NF=1.2dB
                R=.5dB
                Flo=11700MHz
                Fhi=12200MHz

Figure 8-13 Ku band LNA with bandpass filter

Note that a passive device's insertion loss is entered as a negative gain in dB in the Friis formula. The noise figure then becomes the absolute value of this loss or simply 0.2 dB in the case of the bandpass filter. Using the Friis formula to solve for the resulting noise figure leads to the following result.

$$F = 10^{\frac{0.2}{10}} + \frac{10^{\frac{0.7}{10}} - 1}{10^{\frac{-0.2}{10}}} + \frac{10^{\frac{1.1}{10}} - 1}{10^{\frac{-0.2}{10}} \cdot 10^{\frac{15}{10}}} + \frac{10^{\frac{1.2}{10}} - 1}{10^{\frac{-0.2}{10}} \cdot 10^{\frac{15}{10}} \cdot 10^{\frac{12}{10}}} = 1.2402$$

Converting the noise factor back to noise figure gives a total cascaded noise figure of 0.935 dB.

$$F_{dB} = 10 \cdot \log (1.2402) = 0.935 \ dB$$

Based on this simple application of the Friis formula the engineer would expect the system noise figure to be 0.935 dB. Simulating the block diagram of Figure 8-13 in Genesys reveals a vastly different result. The simulated noise figure of Figure 8-14 shows that there is significant ripple in the noise figure. The worst case noise figure is actually greater than 2 dB across the passband of the amplifier. The 0.5 dB ripple in the filter actually results in greater than 1.5 dB ripple in the noise figure. This is due to the fact that the noise figure of the input device is very sensitive to the impedance that is presented to it. The 0.935 dB noise figure is an erroneous result that is obtained when using the Friss formula or one of the many spreadsheet cascade analysis programs.

By modeling the simple cascade in Genesys we can quickly become aware of this condition.

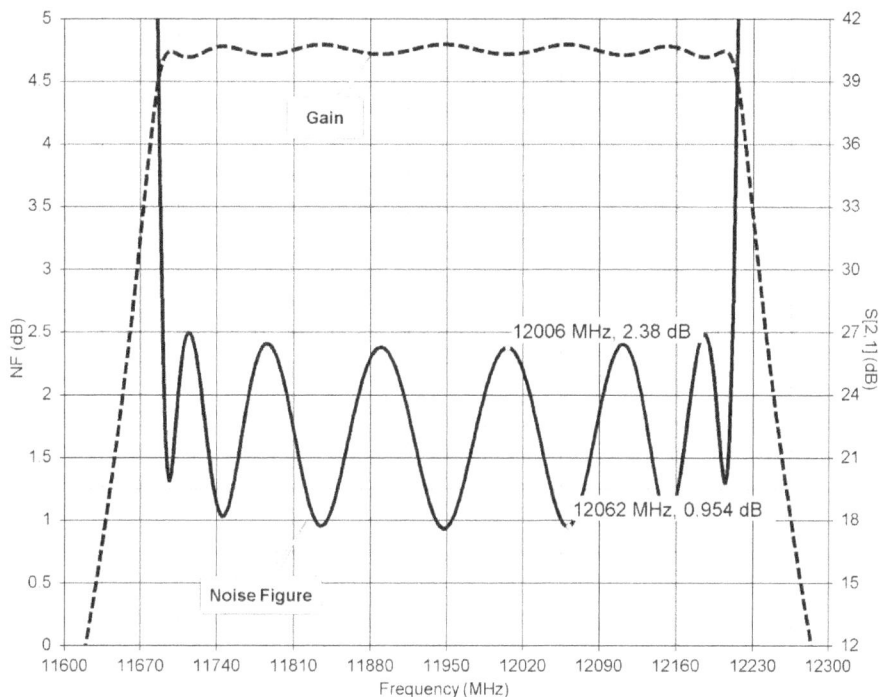

Figure 8-14 Ku band LNA with bandpass filter simulated response

8.12 Reducing the Effect of Source Impedance

In practice it is often desirable to place a low-loss isolator at the input of the LNA to buffer the effects of impedance variation due to the filter ripple. It is important that the isolator have very low loss as its insertion loss will also add to the overall noise figure. This can easily be modeled in Genesys by adding an isolator to the block diagram as shown in Figure 8-15.

Example 8-8: Add an isolator with an insertion loss of 0.1 dB to the block diagram of Figure 8-14 and plot the LNA cascade insertion loss and noise figure from 11.6 to 12.3 GHz

Solution: An isolator with 0.1 dB insertion loss reduces the noise figure by 0.8 dB. Perform a Linear Analysis over the range of 11 GHz to 13 GHz with 600 points. This puts the noise figure of the LNA closer to 1.5 dB, which is a better value for the reception of Ku Band satellite signals from space. The resulting cascaded gain and noise figure is shown in Figure 8-16. Note the tremendous decrease in the noise figure due to the addition of the isolator between the filter and the amplifier.

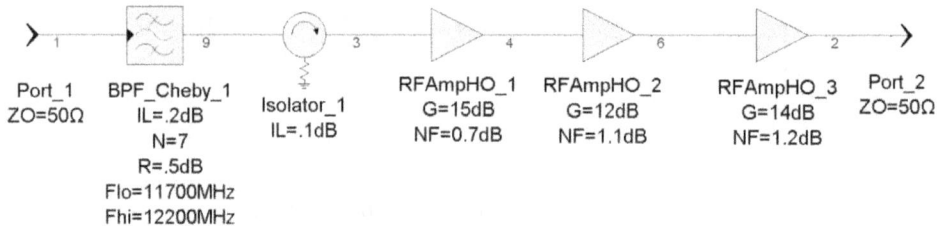

Port_1   BPF_Cheby_1   Isolator_1   RFAmpHO_1   RFAmpHO_2   RFAmpHO_3   Port_2
ZO=50Ω   IL=.2dB   IL=.1dB   G=15dB   G=12dB   G=14dB   ZO=50Ω
              N=7   NF=0.7dB   NF=1.1dB   NF=1.2dB
              R=.5dB
              Flo=11700MHz
              Fhi=12200MHz

Figure 8-15 Ku band LNA with isolator and bandpass filter

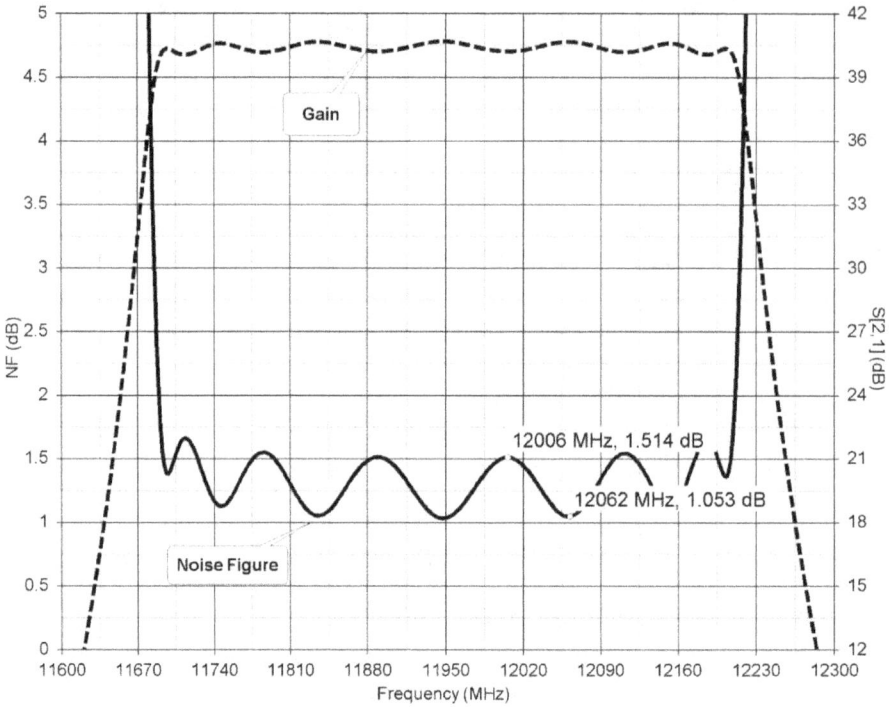

Figure 8-16 LNA cascade noise figure with input matching circuit

8.13 Summary

The example of Section 8.6 gives an introduction to the important subject of system, or block diagram, level simulation. As the previous example shows there are also system level computations that can be evaluated with Linear simulation techniques. Linear simulation continues to be a very important topic and is the foundation for all RF and microwave CAD work. This volume provides the reader with a thorough coverage of the linear circuit design techniques that can be accomplished with Linear simulation in Genesys. As an applications manual this text forms a bridge between the classic theory and practical engineering problem solving.

References and Further Reading

[1]   Guillermo Gonzales, *Microwave Transistor Amplifiers – Analysis and Design,* Second Edition, Prentice Hall Inc., Upper Saddle River, New Jersey

[2]   Randy Rhea, *The Yin-Yang of Matching: Part 1 – Basic Matching Concepts*, High Frequency Electronics, March 2006

[3]   Steve C. Cripps, *RF Power Amplifiers for Wireless Communications*, Artech House Publishers, Norwood, MA. 1999

[4]   David M. Pozar, *Microwave Engineering*, Fourth Edition, John Wiley & Sons, New York, 2012

[5]   R. Ludwig, P. Bretchko, *RF Circuit Design*, Theory and Applications, Prentice Hall, Upper Saddle River, NJ, 2000

[6]   Ali A. Behagi and Stephen D. Turner, *Microwave and RF Engineering,* A Simulation Approach with Keysight Genesys Software, BT Microwave LLC, March, 2015

Problems

8-1. Design a Two–Stage Maximum Gain Amplifier

(a) Use the Agilent HBFP0405 and the Mitsubishi MGF0911a transistors to design a two-stage maximum gain amplifier at 2.2 GHz. All the input, output, and interstage impedance matching networks should be analytically designed using two-element L-networks.

(b) Design the stabilizing network for the first stage using Agilent HBFP0405 transistor

(c) Using the HBFP0405 stabilized transistor create a tabular output of the simultaneous match source and load impedance and Gmax

(d) Design the first stage input matching network by matching the 50 Ohm source impedance to the conjugate impedance looking into the stabilized HBFP0405 device, ZM1.

(e) Design the stabilizing network for the second stage using the Mitsubishi device

(f) Using the MGF0911a stabilized transistor create a tabular output of the simultaneous match source and load impedance and Gmax

(f) Design the second stage output matching network by matching the 50 Ohm load impedance to the conjugate impedance looking into the output of the stabilized MGF0911a device, ZM2

(g) Design the interstage matching network between stage one ZM2 and stage two ZM1.

(h) Cascade the two amplifiers and display the overall response of the cascaded network. The cascaded gain of the amplifier must be the sum of the maximum gains of the two single-stage amplifiers.

8-2. Increase the Bandwidth of the Matching Networks

(a) Increase the bandwidth of the input matching network by selecting a virtual resistor R_M between the input resistor of the matching network, R_S, and the output resistor of the matching network, R_L, such that $R_M = \sqrt{R_S R_L}$ and then conjugately match the input and output impedances to the virtual resistor R_M with L-networks.

(b) Cascade the two matching networks and measure the fractional bandwidth at 20 dB return loss.

(c) Compare the measured fractional bandwidth with the fractional bandwidth of the original matching network at 20 dB return loss and calculate the percentage of the increase in the bandwidth.

(d) Repeat the above procedure for the interstage and output matching networks.

8-3. LNA Cascade Using the Friis Formula

(a) Consider a typical Low Noise Amplifier used in the X Band of 8.7 to 9.2 GHz. Using the Friis formula calculate the overall noise figure of the following LNA cascade.

LNA #1	G1 = 15 dB	F1 = 1.1 dB
LNA #2	G1 = 14 dB	F1 = 1.2 dB
LNA #3	G1 = 13 dB	F1 = 1.3 dB

(This page is intentionally left blank)

Appendix A

Straight Wire Parameters for Solid Copper Wire

Wire Size (AWG)	Diameter in Mils	Resistance Ohms/1000 ft.	Area in circular Mils	Suggested Maximum Current Handling, Amperes[1]
0000	460.0	0.049	211600	1000
000	409.6	0.062	167800	839
00	364.8	0.078	133100	665
0	324.9	0.098	105500	527
1	289.3	0.124	83690	418
2	257.6	0.156	66360	332
3	229.4	0.197	52620	263
4	204.3	0.249	41740	208
5	181.9	0.313	33090	165
6	162.0	0.395	26240	131
7	144.3	0.498	20820	104
8	128.5	0.628	16510	83
9	114.4	0.793	13090	65
10	101.9	0.999	10380	52
11	90.7	1.26	8230	41
12	80.8	1.56	6530	32
13	72.0	2.00	5180	26
14	64.1	2.52	4110	20
15	57.1	3.18	3260	16
16	50.8	4.02	2580	13
17	45.3	5.05	2050	10
18	40.3	6.39	1620	8.0
19	35.9	8.05	1290	6.0
20	32.0	10.1	1020	5.0
21	28.5	12.8	812	4.0
22	25.3	16.2	640	3.0
23	22.6	20.3	511	2.5
24	20.1	25.7	404	2.0
25	17.9	32.4	320	1.6
26	15.9	41.0	253	1.2
27	14.2	51.4	202	1.0
28	12.6	65.3	159	0.80
29	11.3	81.2	123	0.61
30	10.0	104.0	100	0.50
31	8.9	131	79.2	0.40
32	8.0	162	64.0	0.32
33	7.1	206	50.4	0.25
34	6.3	261	39.7	0.19
35	5.6	331	31.4	0.16
36	5.0	415	25.0	0.12
37	4.5	512	20.2	0.10
38	4.0	648	16.0	0.08
39	3.5	847	12.2	0.06
40	3.1	1080	9.61	0.05
41	2.8	1320	7.84	0.04
42	2.5	1660	6.25	0.03
43	2.2	2140	4.84	0.024
44	2.0	2590	4.00	0.020
45	1.76	3350	3.10	0.016
46	1.57	4210	2.46	0.012
47	1.40	5290	1.96	0.010
48	1.24	6750	1.54	0.008
49	1.11	8420	1.23	0.006
50	0.99	10600	0.98	0.005

Appendix B

B-1 Impedance Matching Network Design

One of the important tasks in RF and microwave engineering is the determination of how an arbitrary complex load impedance, $Z_L = R_L + jX_L$, is analytically matched to any complex source impedance, $Z_S = R_S + jX_S$. This problem arises mainly in the design of inter-stage matching networks between active devices or between an antenna and a transmitter where both impedances are usually complex.

In Figure B-1 the complex load impedance, $Z_L = R_L + j X_L$, is to be matched to the source impedance $Z_S = R_S + jX_S$. The only condition for impedance matching is that both R_S and R_L must be nonnegative while X_S and X_L could take any real value. In the *RF and Microwave Circuit Design* textbook it was proved that the maximum power is transferred from the source to the load when the load impedance is the conjugate of the source impedance.

In designing impedance matching networks using only 2 lumped elements, such as L-networks, there are two configurations that can match an arbitrary load impedance to any arbitrary source impedance. In the first configuration the first element adjacent to the load is a series element and the second element is a shunt element. Such configuration is shown in Figure B-1.

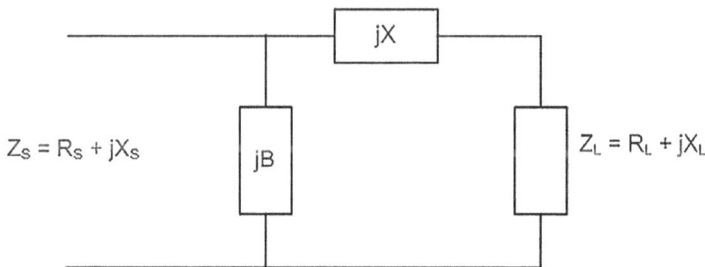

Figure B-1 First impedance matching network configuration

In the second configuration the first element adjacent to the load is a shunt element and the second element is a series element, as shown in Figure B-2.

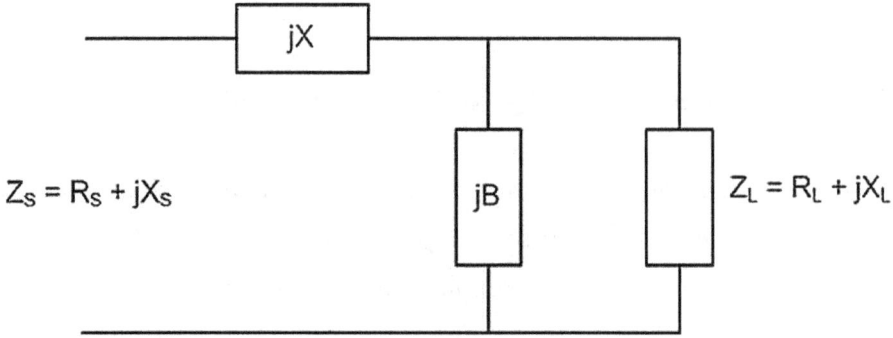

Figure B-2 Second impedance matching network configuration

B-2 Matching a Complex Load to a Complex Source Impedance

To match a complex load to any complex source impedance using a single L-network, either the first or the second configuration may be used. The choice of configurations depends on the conditions that source and load impedances dictate. Applying the maximum power transfer condition $Z_S^* = Z_{IN}$ to the first matching configuration in Figure B-2 we get:

$$R_S - jX_S = \cfrac{1}{jB + \left(\cfrac{1}{jX + R_L \quad jX_L} \right)} \tag{5-6}$$

By separating the real and imaginary parts of Equation (5-6) we obtain two solutions for B and X as follows:

$$B_1 = \frac{R_L X_S \quad \sqrt{R_L R_S (R_S^2 \quad X_S^2 - R_L R_S)}}{R_L (R_S^2 + X_S^2)} \tag{5-7}$$

$$X_1 = \frac{R_L X_S - R_S X_L}{R_S} + \frac{R_S - R_L}{B_1 R_S} \tag{5-8}$$

And

$$B_2 = \frac{R_L X_S - \sqrt{R_L R_S (R_S{}^2 + X_S{}^2 - R_L R_S)}}{R_L (R_S{}^2 + X_S{}^2)} \qquad (5\text{-}9)$$

$$X_2 = \frac{R_L X_S - R_S X_L}{R_S} + \frac{R_S - R_L}{B_2 R_S} \qquad (5\text{-}10)$$

Similarly, applying the same procedure to the second configuration we have:

$$R_S - jX_S = jX + \frac{1}{jB + \left(\dfrac{1}{R_L + jX_L}\right)} \qquad (5\text{-}11)$$

Separating the real and imaginary parts of Equation (5-11), we also get two sets of solutions for B and X:

$$B_3 = \frac{R_S X_L \quad \sqrt{R_L R_S (R_L{}^2 \quad X_L{}^2 - R_L R_S)}}{R_S (R_L{}^2 + X_L{}^2)} \qquad (5\text{-}12)$$

$$X_3 = \frac{R_S X_L - R_L X_S}{R_L} + \frac{R_L - R_S}{B_3 R_L} \qquad (5\text{-}13)$$

And

$$B_4 = \frac{R_S X_L - \sqrt{R_L R_S (R_L{}^2 + X_L{}^2 - R_L R_S)}}{R_S (R_L{}^2 + X_L{}^2)} \qquad (5\text{-}14)$$

$$X_4 = \frac{R_S X_L - R_L X_S}{R_L} + \frac{R_L - R_S}{B_4 R_L} \qquad (5\text{-}15)$$

Conditions for the validity of solutions are that the arguments of the square roots in Equations (5-7), (5-9), (5-12) and (5-14) be positive or zero.

Therefore:

- If $R_S^2 + X_S^2 - R_L R_S > 0$ and $R_L^2 + X_L^2 - R_L R_S < 0$, the two solutions in Equations (5-7) through (5-10) are the only valid solutions.

- If $R_S^2 + X_S^2 - R_L R_S < 0$ and $R_L^2 + X_L^2 - R_L R_S > 0$, the two solutions in Equations (5-12) through (5-15) are the only valid solutions.

- If $R_S^2 + X_S^2 - R_L R_S > 0$ and $R_L^2 + X_L^2 - R_L R_S > 0$, all four solutions in Equations (5-7) through (5-10) and (5-12) through (5-15) are valid.

Once the real values for B and X are calculated, the values of the matching elements are obtained from the following equations:

If B is positive, the matching element is a capacitor given by:

$$C = \frac{B}{2\pi f} \qquad (5\text{-}16)$$

If B is negative, the matching element is an inductor given by:

$$L = -\frac{1}{2\pi f B} \qquad (5\text{-}17)$$

If X is positive, the matching element is an inductor given by:

$$L = \frac{X}{2\pi f} \qquad (5\text{-}18)$$

If X is negative, the matching element is a capacitor given by:

$$C = -\frac{1}{2\pi f X} \qquad (5\text{-}19)$$

In the above equations, if the frequency is in Hz, capacitor and inductor values are in Farad and Henry, respectively.

In practice, we utilize Equations (5-7) through (5-15) to design the matching L-networks.

B-3 Matching Complex Load to Real Source Impedance

In single stage amplifier design a common matching problem is the matching of a complex load impedance, $Z_L = R_L + jX_L$ to real source impedance, $Z_S = R_S$. The complex impedance is usually the load and the real impedance is the characteristic impedance of the transmission line connected to the source. For the case of real source impedance the matching configuration of Figure B-1 is redrawn in Figure B-3.

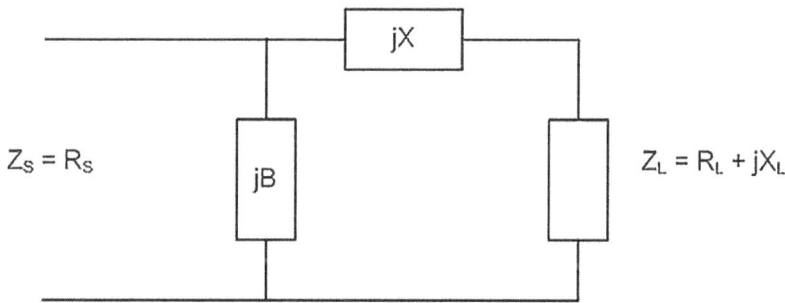

Figure B-3 Matching a resistive source to a complex load impedance

To derive the analytical expressions for B and X we utilize the maximum power transfer condition and set the conjugate of the source impedance equal to input impedance of the matching network followed by the load impedance, as given in Equation (5-20).

$$R_S = \cfrac{1}{jB + \left(\cfrac{1}{jX + R_L + jX_L} \right)} \qquad (5\text{-}20)$$

Note that the conjugate of a real source resistor is equal to itself.

The solutions for B and X in Equation (5-20) can easily be obtained by substituting $X_S = 0$ in Equations (5-7) through (5-10).

$$B_1 = \frac{+\sqrt{R_S - R_L}}{R_S\sqrt{R_L}}$$

(5-21)

$$X_1 = +\sqrt{R_L(R_S - R_L)} - X_L$$

(5-22)

And,

$$B_2 = \frac{-\sqrt{R_S - R_L}}{R_S\sqrt{R_L}}$$

(5-23)

$$X_2 = -\sqrt{R_L(R_S - R_L)} - X_L$$

(5-24)

Note that the solutions given in Equations (5-21) through (5-24) are only valid if $R_L < R_S$. To calculate B and X, when $R_L > R_S$, we use the second matching configuration shown in Figure B-4,

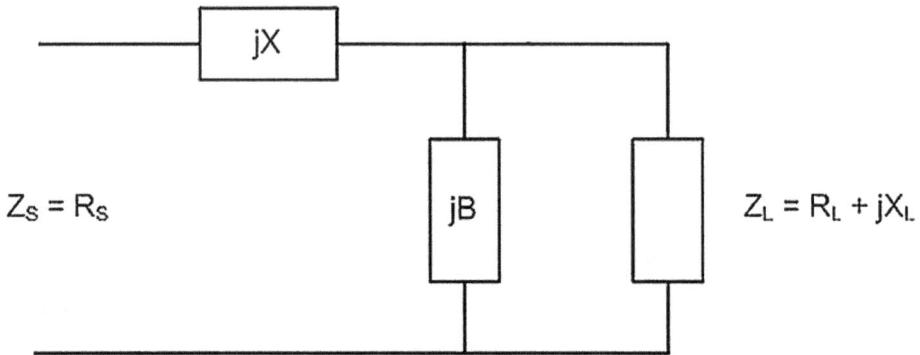

Figure B-4 Matching complex load to resistive source (2nd configuration)

Applying the maximum power condition we have:

$$R_S = jX + \frac{1}{jB + \left(\dfrac{1}{R_L + jX_L}\right)}$$

(5-25)

The solutions for B and X in Equation (5-25) can be obtained by reusing Equations (5-12) and (5-14) and substituting $X_S = 0$ in Equations (5-13) and (5-15).

$$B_3 = \frac{R_S X_L \quad \sqrt{R_L R_S (R_L^2 + X_L^2 - R_L R_S)}}{R_S (R_L^2 + X_L^2)} \tag{5-26}$$

$$X_3 = \frac{R_S X_L}{R_L} + \frac{R_L - R_S}{B_3 R_L} \tag{5-27}$$

And

$$B_4 = \frac{R_S X_L - \sqrt{R_L R_S (R_L^2 + X_L^2 - R_L R_S)}}{R_S (R_L^2 + X_L^2)} \tag{5-28}$$

$$X_4 = \frac{R_S X_L}{R_L} + \frac{R_L - R_S}{B_4 R_L} \tag{5-29}$$

The conditions for the valid solutions are that the arguments of the square roots in Equations (5-26) through (5-29) be non-negative. Therefore, the two solutions obtained from Equations (5-26) through (5-29) are valid only if $R_L > R_S$. The combined conditions are summarized in Table B-1.

Case #	First Condition	Second Condition	# of Solutions	Equations Used
1	$R_L < R_S$	$R_L^2 + X_L^2 - R_L R_S > 0$	4	(5-21) to (5-24) (5-26) to (5-29)
2	$R_L < R_S$	$R_L^2 + X_L^2 - R_L R_S < 0$	2	(5-21) to (5-24)
3	$R_L > R_S$	N/A	2	(5-26) to (5-29)

Table B-1 Impedance matching conditions and the number of solutions

The solutions in Equations (5-26) through (5-29) can be simplified by normalizing the load impedance with respect to the source resistor. Therefore, if $R_S = Z_0$, the normalized load resistance is:

$$r = \frac{R_L}{Z_0} \tag{5-30}$$

And

$$x = \frac{X_L}{Z_0} \tag{5-31}$$

The simplified equations for matching a complex load to a real source are:

$$B_1 = \frac{\sqrt{\frac{(1-r)}{r}}}{Z_0} \tag{5-32}$$

$$X_1 = Z_0 \left[\sqrt{r(1-r)} - x \right] \tag{5-33}$$

$$B_2 = -\frac{\sqrt{(1-r)}}{Z_0} \tag{5-34}$$

$$X_2 = -Z_0 \left[\sqrt{r(1-r)} + x \right] \tag{5-35}$$

$$B_3 = \frac{x \sqrt{r(r^2 + x^2 - r)}}{Z_0 (r^2 + x^2)} \tag{5-36}$$

$$X_3 = Z_0 \sqrt{\frac{(r^2 + x^2 - r)}{r}} \tag{5-37}$$

$$B_4 = \frac{x - \sqrt{r\left(r^2 + x^2 - r\right)}}{Z_0\left(r^2 + x^2\right)} \qquad (5\text{-}38)$$

$$X_4 = -Z_0\sqrt{\frac{\left(r^2 + x^2 - r\right)}{r}} \qquad (5\text{-}39)$$

B-4 Matching a Real Load to a Real Source Impedance

When source and load impedances are both real, the first matching configuration can be applied.

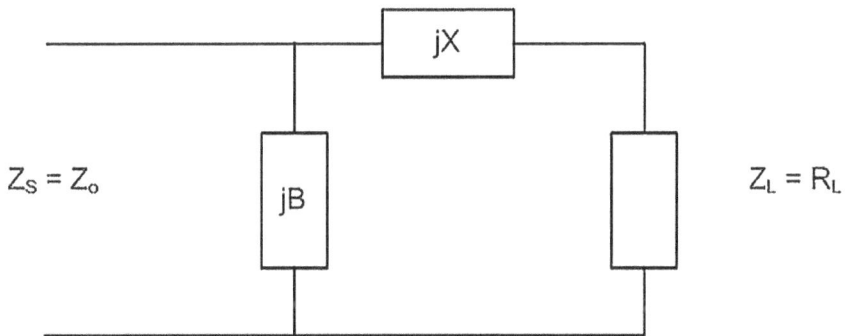

Figure B-5 First matching configuration with $X_L = X_S = 0$

To design the matching network, apply the maximum power transfer condition and require that $Z_0^* = Z_{IN}$. Therefore,

$$Z_0 = \frac{1}{jB + \left(\dfrac{1}{jX + R_L}\right)} \qquad (5\text{-}40)$$

Substituting $r = \dfrac{R_L}{Z_0}$ in Equation (5-40), we get:

$$Z_0 = \frac{1}{jB + \left(\dfrac{1}{jX + rZ_0}\right)} \tag{5-41}$$

The solutions for B and X can be obtained by using Equations (5-32) and (5-34) and substituting $x = 0$ in Equations (5-33) and (5-35).

$$B_1 = \frac{\sqrt{(1-r)/r}}{Z_0} \tag{5-42}$$

$$X_1 = Z_0\sqrt{r(1-r)} \tag{5-43}$$

$$B_2 = -\frac{\sqrt{(1-r)/r}}{Z_0} \tag{5-44}$$

$$X_2 = -Z_0\sqrt{r(1-r)} \tag{5-45}$$

Note that the two solutions given by Equations (5-42) through (5-45) are only valid if r is less than 1 or $R_L < Z_0$.

If $r > 1$, we use the second matching configuration as in Figure B-6.

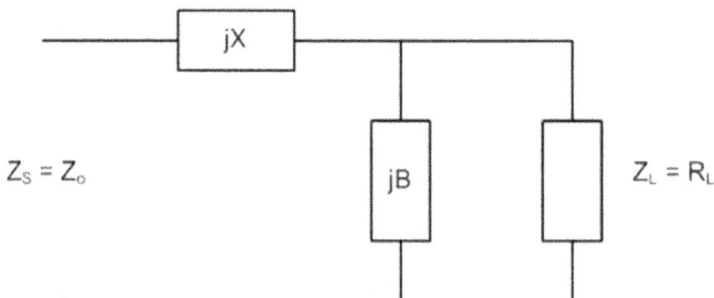

Figure B-6 Second matching configuration for resistive load and source

To calculate the B and X values, we require that, $Z_0{}^* = Z_{IN}$, therefore,

$$Z_0 = jX + \cfrac{1}{jB + \left(\cfrac{1}{Z_0 r}\right)}$$

(5-46)

The solutions for B and X in Equation (5-46) can be obtained by substituting x = 0 in Equations (5-36) through (5-39).

$$B_3 = \frac{\sqrt{r-1}}{Z_0 r}$$

(5-47)

$$X_3 = Z_0 \sqrt{r-1}$$

(5-48)

$$B_4 = \frac{-\sqrt{r-1}}{Z_0 r}$$

(5-49)

$$X_4 = -Z_0 \sqrt{r-1}$$

(5-50)

Note that the two solutions given by Equations (5-47) through (5-50) are only valid if r is greater than 1 or $R_L > Z_0$.

The combined conditions for the validity of solutions are given below.

Case 1 If r is less than 1 there are two L-networks that match the two impedances. The two solutions are given by Equations (5-42) through (5-45).

Case 2 If r is greater than 1 there are two L-networks that match the two impedances. The two solutions are given by Equations (5-47) through (5-50).

Appendix C

C-1 Analytical Design of the Quarter-Wave Matching Networks

We have shown that the input impedance of a quarter-wave network with characteristic impedance, Z_0, terminated in a resistive load, R_L, is given by:

$$Z_{IN} = \frac{Z_0^{\,2}}{R_L}$$

This equation can be written as:

$$Z_O = \sqrt{R_L Z_{IN}} \tag{6-1}$$

We have also shown that the maximum power transfer from a source with resistance R_S to a network terminated in a resistive load R_L is achieved only if the input impedance of the network is equal to the source resistance.

$$Z_{IN} = R_S$$

Therefore, for maximum power transfer, the characteristic impedance of the quarter-wave network must satisfy the Equation (6-2).

$$Z_O = \sqrt{R_S R_L} \tag{6-2}$$

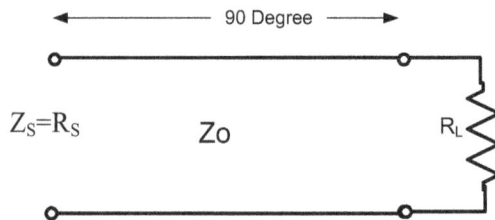

Figure C-1 Quarter-wave network terminated in resistor R_L

Equation (6-2) states that the characteristic impedance of the quarter-wave network, matching R_L to R_S, must be equal to the square root of the product of source and load resistors.

If we normalize the load resistor R_L with respect to R_S,

$$r = \frac{R_L}{R_S} \tag{6-3}$$

Equation (6-2) can be written as Equation (6-4).

$$Z_O = R_S \sqrt{r} \tag{6-4}$$

Notice that for $R_L > R_S$, the characteristic impedance Z_0 is greater than R_S while for $R_L < R_S$, the characteristic impedance Z_0 is less than R_S.

The fractional bandwidth of a network, *FBW*, is defined in Equation (6-5) where f_H and f_L are the upper and lower frequencies of the bandwidth and the center frequency f_0 is equal to $\sqrt{f_H f_L}$, respectively.

$$FBW = \frac{f_H - f_L}{\sqrt{f_H f_L}} \tag{6-5}$$

The fractional bandwidth of a quarter-wave matching network is given in Equation (6-6):

$$FBW = 2 - \frac{4}{\pi} \cdot \cos^{-1}\left(\frac{2\Gamma_m \sqrt{r}}{\sqrt{1 - \Gamma_m^2}|1 - r|} \right) \tag{6-6}$$

Where, Γ_m is the magnitude of the reflection coefficient.

Equation (6-6) shows that the fractional bandwidth of a quarter-wave matching network depends on the magnitude of the input reflection coefficient, Γ_m, and the mismatch ratio, r. The solution to Equation (6-6) is only valid if,

$$\frac{2\Gamma_m \sqrt{r}}{\sqrt{1-\Gamma_m^2}\,|1-r|} \leq 1$$

At 3 dB return loss the reflection coefficient is $\Gamma_m = 0.707$, therefore, Equation (6-6) reduces to:

$$FBW_{3dB} = 2 - \frac{4}{\pi} \cdot \cos^{-1}\left(\frac{2\sqrt{r}}{|1-r|}\right) \tag{6-7}$$

At 3 dB return loss Equation (6-7) has valid solutions only if,

$$\frac{2\sqrt{r}}{|1-r|} \leq 1 \quad or \quad r^2 - 6r + 1 \geq 0 \tag{6-8}$$

Similarly, if we define the bandwidth at $\Gamma_m = 0.1$, corresponding to 20 dB return loss, as a good matching bandwidth it is insightful to evaluate the fractional bandwidth associated with this 20 dB return loss. From Equation (6-6) the fractional bandwidth at $\Gamma_{in} = 0.1$, is:

$$FBW_{20dB} = 2 - \frac{4}{\pi} \cdot \cos^{-1}\left(\frac{0.2\sqrt{r}}{\sqrt{0.99}\,|1-r|}\right) \tag{6-9}$$

At 20 dB return loss Equation (6-9) has valid solutions only if,

$$\frac{0.2\sqrt{r}}{\sqrt{0.99}\,|1-r|} \leq 1 \quad or \quad 99r^2 - 202r + 99 \geq 0 \tag{6-10}$$

The loaded quality factor, Q_L, of the quarter-wave matching network is defined as the inverse of the fractional bandwidth at 3 dB return loss; therefore, the loaded Q factor can be calculated from Equation (6-11).

$$Q_L = \frac{1}{FBW_{3dB}} = \frac{1}{2 - \frac{4}{\pi} \cdot \cos^{-1}\left(\frac{2\sqrt{r}}{|1-r|}\right)} \tag{6-11}$$

Equation (6-11) shows that the validity condition in Equation (6-8) for the 3 dB fractional bandwidth is the same for the Q factor except that whenever the 3 dB fractional bandwidth tends towards infinity the Q factor tends towards zero. The Q factor given in Equation (6-11) is not to be confused with the unloaded transmission line Q factor. This is actually an external Q factor as it relates to the loaded Q of the overall network.

If f is the center (design) frequency, the bandwidth of the circuit is then calculated from Equation (6-12).

$$BW = (f) \cdot (FBW) \tag{6-12}$$

C-2 Analytical Design of the Series Transmission Line

It has been shown in the *RF and Microwave Circuit Design* textbook that the input impedance of a lossless transmission line of length d and characteristic impedance Z_0 terminated in an arbitrary load Z_L, is given in the following equation.

$$Z_{IN} = Z_o \frac{Z_L \quad jZ_o \tan \beta d}{Z_o + jZ_L \tan \beta d}$$

Setting $Z_0 = R_S$ and $\tan \beta d = t$, the input admittance of the network in Figure C-1 can be written as:

$$Y_{IN} = \frac{1}{Z_{IN}} = \frac{R_S + jZ_L t}{R_S(Z_L + jR_S t)} = G_{IN} + jB_{IN} \tag{6-13}$$

Substituting the normalized load impedance, $Z_L/R_s = r + jx$ into Equation (6-13) and separating its real and imaginary parts we get:

$$G_{IN} = \frac{r(1+t^2)}{R_S(r^2 + x^2 + t^2 + 2xt)} \tag{6-14}$$

And,

$$B_{IN} = \frac{xt^2 + (r^2 + x^2 - 1)t + x}{R_S(r^2 + x^2 + t^2 + 2xt)} \tag{6-15}$$

The value of d, which implies t, can be obtained by setting the input conductance, G_{IN}, equal to source conductance:

$$\frac{r(1+t^2)}{R_S(r^2 + x^2 + t^2 + 2xt)} = \frac{1}{R_S} \tag{6-16}$$

Equation (6-16) can be rearranged as:

$$(r-1) \cdot t^2 - 2xt - \left(r^2 + x^2 - r\right) = 0 \tag{6-17}$$

Notice that the quadratic Equation (6-17) has two solutions for t. For a wider bandwidth and lower loss usually the smaller value of t is selected. The two solutions for t are Equations (6-180 and (6-19):

$$t_1 = \frac{x + \sqrt{r(r^2 + x^2 - 2r + 1)}}{r - 1} \tag{6-18}$$

And,

$$t_2 = \frac{x - \sqrt{r(r^2 + x^2 - 2r + 1)}}{r - 1} \qquad (6\text{-}19)$$

With $\tan \beta d = t$, and $\beta \lambda = 2\pi$, we have $d = \frac{\lambda}{2\pi} \tan^{-1} t$ and the two solutions

for d are:

$$d_1 = \frac{\lambda}{2\pi} \tan^{-1} t_1 \qquad t_1 \geq 0 \qquad (6\text{-}20)$$

$$d_2 = \frac{\lambda}{2\pi} \tan^{-1} t_2 \qquad t_2 \geq 0 \qquad (6\text{-}21)$$

To specify the lengths of d_1 and d_2 in electrical degrees, we get:

$$d_1 = \frac{360}{2\pi} \tan^{-1} t_1 \qquad t_1 \geq 0 \qquad (6\text{-}22)$$

$$d_2 = \frac{360}{2\pi} \tan^{-1} t_2 \qquad t_2 \geq 0 \qquad (6\text{-}23)$$

Because at every half wavelength the input impedance of a transmission line repeats, there are an infinite number of transmission line lengths that matches the load to source impedance. Usually the shorter length is selected to improve the matching bandwidth. If t_1 or t_2 is negative, we add half a wavelength to each line to get positive d_1 and d_2.

$$d_1 = \frac{360(\pi + \tan^{-1} t_1)}{2\pi} \qquad t_1 < 0 \qquad (6\text{-}24)$$

$$d_2 = \frac{360(\pi + \tan^{-1} t_2)}{2\pi} \qquad t_2 < 0 \qquad (6\text{-}25)$$

C-3 Analytical Design of Shunt Stub Matching Networks

To calculate the electrical length of the shunt stub, first substitute t_1 and t_2 in Equation (6-15) to determine B_1 and B_2.

$$B_1 = \frac{xt_1^2 \quad (r^2 + x^2 - 1)t_1 + x}{R_S(r^2 + x^2 + t_1^2 + 2xt_1)} \tag{6-26}$$

$$B_2 = \frac{xt_2^2 \quad (r^2 + x^2 - 1)t_2 + x}{R_S(r^2 + x^2 + t_2^2 + 2xt_2)} \tag{6-27}$$

Then, the electrical lengths of the open circuited stubs are found by setting the susceptance of the stubs equal to the negative of the input susceptance.

$$so_1 = \frac{-\lambda\left(tan^{-1}(R_S B_1)\right)}{2\pi} \tag{6-28}$$

$$so_2 = \frac{-\lambda\left(tan^{-1}(R_S B_2)\right)}{2\pi} \tag{6-29}$$

If either stub length in Equations (6-28) or (6-29) is negative, add one half wavelength to obtain a positive stub length. For short-circuited stubs, the two solutions are:

$$ss_1 = \frac{\lambda\left(tan^{-1}\left(\frac{1}{R_S B_1}\right)\right)}{2\pi} \tag{6-30}$$

$$ss_2 = \frac{\lambda\left(tan^{-1}\left(\frac{1}{R_S B_2}\right)\right)}{2\pi} \tag{6-31}$$

About the Author

Ali Behagi received the Ph.D. degree in electrical engineering from the University of Southern California and the MS degree in electrical engineering from the University of Michigan. He has several years of industrial experience with Hughes Aircraft and Beckman Instruments.

Dr. Behagi joined Penn State University as an associate professor of electrical engineering in 1986. He has devoted over 20 years to teaching RF and microwave engineering courses and directing university research projects. While at Penn State he received the National Science Foundation grant to establish a microwave and RF engineering lab. He also received the Keysight/Agilent software grant to use in teaching high frequency circuit design courses and laboratory experiments. In the past few years he has been very active in authoring RF and microwave circuit design textbooks with 100 design examples using the Keysight Genesys and ADS software.

Dr. Behagi is a Keysight Distinguished Author and a Certified Expert. He is also a Lifetime Member of the Institute of Electrical and Electronics Engineers (IEEE), and the Microwave Theory and Techniques Society.

www.ingramcontent.com/pod-product-compliance
Lightning Source LLC
Chambersburg PA
CBHW082004190326
41458CB00010B/3072